Sofie Amelie Burger

High Cycle Fatigue of Al and Cu Thin Films by a Novel High-Throughput Method

Schriftenreihe
des Instituts für Angewandte Materialien
Band 23

Karlsruher Institut für Technologie (KIT)
Institut für Angewandte Materialien (IAM)

Eine Übersicht über alle bisher in dieser Schriftenreihe erschienenen Bände finden Sie am Ende des Buches.

High Cycle Fatigue of Al and Cu Thin Films by a Novel High-Throughput Method

by
Sofie Amelie Burger

Dissertation, Karlsruher Institut für Technologie (KIT)
Fakultät für Maschinenbau
Tag der mündlichen Prüfung: 04. Dezember 2012

Impressum

Karlsruher Institut für Technologie (KIT)
KIT Scientific Publishing
Straße am Forum 2
D-76131 Karlsruhe
www.ksp.kit.edu

KIT – Universität des Landes Baden-Württemberg und
nationales Forschungszentrum in der Helmholtz-Gemeinschaft

KIT Scientific Publishing 2013
Print on Demand

ISSN 2192-9963
ISBN 978-3-7315-0025-4

High Cycle Fatigue of Al and Cu Thin Films by a Novel High-Throughput Method

Zur Erlangung des akademischen Grades

Doktor der Ingenieurwissenschaften

der Fakultät Maschinenbau

Karlsruher Institut für Technologie (KIT)

genehmigte

Dissertation

von

Dipl.-Ing. Sofie Amelie Burger

aus Stuttgart

Tag der mündlichen Prüfung:	04. Dezember 2012
Hauptreferent:	Prof. Dr. Oliver Kraft
Korreferent:	Prof. Dr. Alexander Wanner

Für meine Eltern

Wissenschaft ist Irrtum auf den letzten Stand gebracht.

Linus Pauling

Danksagung

Die vorliegende Arbeit wurde im Zeitraum von September 2008 bis September 2012 am Institut für Angewandte Materialien am Karlsruher Institut für Technologie angefertigt. An dieser Stelle möchte ich mich bei allen bedanken, die direkt oder indirekt zum Gelingen dieser Arbeit beigetragen haben.

Mein herzlicher Dank gilt Dr. Christoph Eberl für seine engagierte Betreuung meiner Arbeit. Chris, Deine stets positive Art und Weise haben sehr geholfen, auch in etwas frustrierenden wissenschaftlichen Momenten, nicht den Mut zu verlieren, dass doch etwas funktionieren kann, obwohl es auf den ersten Augenblick nicht so scheint.

Prof. Oliver Kraft danke ich für die Übernahme des Hauptreferats, sowie dass er mir die Möglichkeit gegeben hat dieses spannende Thema am Institut zu bearbeiten. Prof. Alexander Wanner danke ich für die freundliche Übernahme des Korreferats.

Herrn Alexander Siegel und Prof. Alfred Ludwig von der Abteilung Werkstoffe der Mikrotechnik/Institut für Werkstoffe der Ruhr Universität Bochum danke ich für die Herstellung und Beschichtung der Ermüdungsproben.

Mein Dank gilt Herrn Ewald Ernst für seine tatkräftige Unterstützung im Labor und Frau Daniela Exner für die Einweisung am Dual Beam Mikroskop. Matthias Funk und Byung-Gil Yoo danke ich für die Aufnahme von TEM Bildern, sowie Jochen Lohmiller und Patric Gruber für die XRD Messungen.

Ein großes Dankeschön an alle ehemaligen und aktuellen Doktoranden und Kollegen der Abteilung für eine angenehme Arbeitsatmosphäre und auch viele lustige Freizeitunternehmungen. Simi, Tina, Matthias, Sven, Thomas, Christian und Martin: Danke für eine tolle Zeit!

Besonders danken möchte ich meiner Familie und Freunden, die immer für mich da waren. Meinen Eltern widme ich diese Arbeit, weil sie immer an mich geglaubt und mich zu jeder Zeit unterstütz haben. Mama und Papa: Vielen Dank für Alles, Ihr seid wunderbar!

Abstract

In the last two decades, the reliability of small electronic devices used in automotive or consumer electronics gained researchers attention. Thus, there is the need to understand the fatigue properties and damage mechanisms of thin films. In this thesis a novel high-throughput testing method for thin films on Si substrate is presented. The literature review gives insights to small scale materials behavior, which is influenced by sample or microstructural dimensions. Several model perceptions have been established over the past years to explain the exceptional small scale behavior for monotonic and cyclic loading.

For testing thin films and their fatigue properties in a high-throughput manner, a novel resonant bending setup has been developed. Detailed Finite Element simulation as well as the implementation of the fatigue setup is presented. The specialty of this method is the induced strain gradient in the thin film along the cantilever due to bending. This gives the opportunity to test one sample at different strain amplitudes at the same time and measure an entire lifetime curve with only one experiment. To characterize the damage morphology of the fatigued thin films, mainly microscopic methods have been used. It is found that the damage morphology differs depending on the tested material, the initial microstructure, and interfaces. Damage in Al thin films is homogeneously distributed and consists of hillocks and extrusions. Possible mechanisms are mediated by diffusion, grain boundaries, and/or dislocations. The Cu thin films display islands of extrusions caused by dislocation processes, additionally influenced by interfaces. Based on the results a sequence of damage mechanisms is proposed. Furthermore, a new form of lifetime diagram is introduced.

Kurzzusammenfassung

In den letzten beiden Jahrzehnten gewann die Zuverlässigkeit von kleinen, elektronischen Bauteilen in der Automobil- und Unterhaltungsindustrie zunehmend das Interesse der Forschung. Deshalb ist es von großer Wichtigkeit die Ermüdungseigenschaften und Schadensmechanismen, vor allem in dünnen Schichten zu kennen und zu verstehen. In der vorliegenden Arbeit wird eine neue Hochdurchsatz-Methode präsentiert, mit der man dünne Schichten auf Siliziumsubstrat auf ihre Ermüdungseigenschaften testen kann. Der Literaturteil am Anfang, gibt einen Überblick über das mechanische Verhalten von kleinskaligen Materialien, das einerseits von der Probengeometrie und andererseits der Mikrostruktur beeinflusst wird. Verschiedene Modellvorstellungen sind im Laufe der Jahre entwickelt worden, um das besondere kleinskalige Verhalten für monotone und zyklische Belastung zu beschreiben.

Um nun dünne Schichten und ihre Ermüdungseigenschaften im Hochdurchsatz zu testen, ist ein neuer Resonanz-Biegeversuchsstand entwickelt worden. Detaillierte Finite Element Simulationen sind durchgeführt worden und die Implementierung des Ermüdungsversuchsstandes wird erläutert. Das Besondere dieser Methode ist, dass in der dünnen Schicht, auf Grund der Balkenbiegung, ein Dehnungsgradient entlang des Balkens eingebracht wird. Dies wiederum ermöglicht das Bestimmen einer kompletten Wöhler-Kurve mit verschiedenen Dehnungsamplituden in nur einem Experiment. Zur Charakterisierung der Schädigungsmorphologie der ermüdeten dünnen Schichten sind hauptsächlich mikroskopische Methoden verwendet worden. Es wurde herausgefunden, dass die Schädigungsmorphologie vom Material, anfänglicher Mikrostruktur und Grenzflächen abhängt. In den Al-Schichten ist die Schädigung homogen an der Dünnschichtoberfläche verteilt und besteht aus Hügeln und Extrusionen. Mögliche Mechanismen zur Entstehung sind durch Diffusion, Korngrenzen und/oder Versetzungen bestimmt. Die Cu-Schichten zeigen Extrusionsinseln mit unterschiedlichen Formen, die durch Versetzungsprozesse und durch Grenzflächen beeinflusst zustande kommen. Basierend auf diesen Ergebnissen wird eine Sequenz der möglichen Schädigungsmechanismen vorgeschlagen. Zudem wird eine neue Art des Lebensdauerdiagramms vorgestellt.

Table of Contents

Symbols and Abbreviations

A	abscissa for maximum strain amplitude fit
an	annealed
as dep	as deposited
b	fatigue sensitivity exponent
B	slope for maximum strain amplitude fit
bcc	body centered cubic
c	fatigue ductility exponent
C	abscissa for local strain amplitude fit
d	grain size
D	slope for local strain amplitude fit
Δ	difference
DIC	Digital Image Correlation
E	Young's modulus
ε_{a}	strain amplitude
$\varepsilon_{\mathrm{a}}^{\mathrm{loc}}$	local strain amplitude
$\varepsilon_{\mathrm{a}}^{\mathrm{max}}$	maximum strain amplitude
$\varepsilon_{\mathrm{el}}$	elastic strain
ε_{f}	fatigue ductility coefficient
$\varepsilon_{\mathrm{pl}}$	plastic strain
$\varepsilon_{z}^{\mathrm{surf}}$	surface strain in z-direction
f	frequency
f_{res}	resonant frequency
F	fraction of damage
F_{max}	maximum fraction of damage
fcc	face centered cubic
FE	Finite Element
FIB	Focused Ion Beam
HCF	high cycle fatigue
I	second moment of inertia
K	fit parameter for lifetime prediction model
L	length
LCF	low cycle fatigue
m_0	mass of cantilever

m_{att}	mass of attached weight
M_z	bending moment in z-direction
n	reaction mode of lifetime prediction model
N	number of cycles
N_{f}	number of cycles to failure
N_{F}	number of cycles to specific fraction of damage
$N_{0.5}$	half lifetime
nc	nanocrystalline
ν	Poisson's ratio
P_z	normal point load in z-direction
PSB	persistent slip band
q	fit parameter for lifetime prediction model
ρ	density
R	loading ratio
σ_{f}	fatigue strength
σ_z^{surf}	surface stress in z-direction
SAW	surface acoustic wave
SEM	Scanning Electron Microscope
TEM	Transmission Electron Microscope
t_{sub}	substrate thickness
u_z	displacement in z-direction
u_z^{max}	maximum displacement in z-direction
ufg	ultra fine-grained
VHCF	very high cycle fatigue
w	width
x	position in x-direction
XRD	X-Ray Diffraction
y	position in y-direction
z	position in z-direction

1. Introduction

Fatigue names the phenomenon that materials fail after reoccurring loading although the applied stresses are below their tensile strength. Fatigue of metals was first studied in 1829 by W.A.J. Albert, when he repeatedly bent mine-hoist chains made of iron [1]. During the beginning of the industrialization in the 19th century, the investigations on fatigue went on, especially after the continued observation of railway axel failure lacking obvious reasons. When August Wöhler [2] investigated the failure of these axels, he established the characterization of fatigue behavior in terms of stress versus number of cycles to failure (S-N) curves. During the 20th century research on fatigue behavior of different materials evolved quickly and several empiric laws were introduced [3-5] to describe the lifetime at low and high cycle numbers, which are still in use to this day. In the late 1980's, the investigation of thin film materials became popular after it was discovered that sample size has an influence on static materials properties [6] such as flow stress. As research has gone deep into the field and many papers have been published, a good overview of thin film material properties is currently available. While monotonic investigations of thin films have developed quickly over the last 30 years, studies on cyclic behavior have been performed less intensively. In the late 1990's researchers started to explore the fatigue behavior of thin foils (e.g. [7]). The investigation of fatigue properties of thin films became more important in the early 2000's, when Ag, Cu and Al thin films were of interest, especially for electronic devices. New techniques had to be established, as the classical testing procedures would not work for small dimensions. Nevertheless, most methods were limited to investigating thin film fatigue behavior only up to around 10^6 cycles.

Since microelectronics for automotive, aeronautics, and consumer applications develop more and more in the direction of getting smaller to have faster devices with higher computational power, this demands a high reliability of such devices. Thin films or lines are especially important as they have to undergo several millions of mechanical or thermal cycles while maintaining their functionality and environmental sustainability. For a good efficiency of measuring thin film fatigue properties, higher testing frequencies must be used to achieve higher cycle numbers in a reasonable amount of time. Up to now an appropriate method for thin film fatigue testing is still missing.

In this thesis, a novel method of thin film fatigue testing is developed. It uses the principle of an alternating strain gradient which is introduced by cyclic bending of cantilevers in resonance. This method is highly efficient and can be seen as a high-throughput method for thin film fatigue testing, especially for cycle numbers up to 10^9. The investigated samples are pure 1 µm Al and Cu thin films with different microstructures sputter deposited onto a Si substrate. In chapter 2 a literature overview on the topic of thin films, fatigue, and reasons for damage mechanisms at the microscale will be given. The idea of the novel high-throughput setup will be introduced in chapter 3, including Finite Element analysis of the cantilever bending, the implementation of the actual setup, and the possibilities and prospects of the setup for observing lifetime characteristics. Other experimental techniques for thin film deposition and characterization are shown in chapter 4. In chapter 5 the results of cyclic loaded Al and Cu thin films are presented in the form of microstructural and quantitative analysis of the damage formation and the lifetime diagrams. The data will be discussed in chapter 6 with respect to the possible damage mechanisms and their causes. Furthermore, a new form of lifetime diagram and a phenomenological lifetime model is proposed. The thesis is concluded and summarized in chapter 7.

2. Theory

In this chapter, a literature overview will introduce the characteristic properties of small scale samples and thin film materials. Their possible damage mechanisms will be presented. Furthermore, fatigue behavior in general and in thin films will be described.

2.1. Small Scale Samples and Thin Films

Already in 1959 Richard Feynman gave an inspiring talk with the title "There is plenty of room at the bottom" [8]. His vision of micro- and nanostructured devices became true with the introduction of processing routes enabling novel materials properties and applications. The development of new automotive and consumer electronics over the past 50 years was only possible with the reduction of device dimensions. This in turn, opened up a new research field for small scale materials testing. As sample dimensions decrease, it was found that materials follow the wording "smaller is stronger" by Yonggang Huang [9]. The small scale samples showed enhanced materials properties compared to their bulk counterparts. This is because sample dimensions are reduced to the same size as the microstructural dimensions e.g. the grain size. Thus, the materials properties are affected by either geometric scaling or microstructural size effects. A scaling effect that is closely related to thin films is the increasing surface to volume ratio. Therefore, oxidation and diffusion controlled processes, e.g. stress relaxation, take place at a much higher rate. Furthermore, the size effect comes into play when e.g. dislocation nucleation and motion is constrained. Comparative reviews on size effects and their various aspects and appearances are summarized in several papers, for example Arzt [10], Dehm et al. [11], Hemker and Sharpe [12], Kraft et al. [13], or Greer and de Hosson [14].

Thin films are a special case of small scale samples, as the sample thickness is significantly reduced compared to the other two dimensions. Thin film behavior can be observed when the film thickness is on the order of several nm up to some microns. The microstructure of thin films strongly depends on the deposition technique and parameters as well as the interface to the substrate as it is described e.g. by Ohring [15]. A more general overview on thin film properties is given in the textbook by Freund and Suresh [16]. Usually, thin films deposited at low temperatures have a grain size of less than the film thickness. In contrast, thin film deposition at high substrate temperatures will lead to a grain size of roughly the film thickness and often forms a columnar grain structure. Testing techniques for small scale materials properties are presented in the following paragraph.

2.1.1. Experimental Techniques for Small Scale Testing

Small scale and thin film testing is still a challenging task, because fabrication, handling, and measuring load or strain are not as easy as for bulk materials. Several techniques have been established for measuring different materials properties. Nanoindentation [17-19] is widely used for thin film characterization. Hardness or Young's Modulus can be measured as a function of the film thickness. Although, there is a complex stress state underneath the indenter tip, the experiment itself is fairly easy. Nanoindentation is also used to test small volumes such as pillars in compression [20]. Contrary, small scale tensile testing can be accomplished by thin films deposited on a compliant substrate or in form of free-standing thin films and nano whiskers. Tensile tests are often combined with *in situ* techniques like X-Ray Diffraction [21, 22], μ-Laue Diffraction [23], or inside an electron microscope [24, 25] to gain additional microstructural information. For measuring strains in these small samples, Digital Image Correlation (DIC) enables a tool for non contact strain measurements [26, 27] independent of the sample size. Wafer curvature is a popular technique to measure the thermal stress in thin films on substrates while undergoing thermal cycling [6]. Recommended review papers on other testing techniques for small scale samples are for example Vinci and Vlassak [28] and Kraft and Volkert [29].

2.1.2. Deformation Mechanisms and Microstructural Features in Small Scale Materials

There are numerous effects that have an influence on the mechanical properties of small scale samples and thin films. In this paragraph, deformation mechanisms and the role of microstructural features are discussed as well as relevant models, which have either been established within the past years or may still be debated. The focus of this chapter is put on thin film materials.

Dislocations

Dislocations are the main carrier of plastic deformation in metals. Several models were proposed to describe the increased flow stress of thin films with respect to their film thickness or microstructure. One of the first papers on this topic was published by Chaudhari [30], who tried to explain the effect by considering the energy balance of the elastic strain and the dislocation field in a thin film. Murakami [31] came up with the idea that dislocations are pinned at the passivated surface and the interface with the substrate. To drive the dislocation forward, it has to bow out between the pinning points similar to the Orowan mechanism. The flow stress, in this case, scales inversely with the film thickness. Murakami's idea was modified by Nix and Freund [6, 32]. Their model proposes that dislocations from the substrate are threading through a single crystalline thin film, and can only move by creating misfit dislocations at the film substrate interface, which contributes to higher stresses. Dörner and Nix [33] later found that grain size also plays a role for thin film strength and added a Hall-Petch type relation [34, 35] to the model. A further refinement was presented by Thompson [36]. He took into account the contribution of dislocations laying down a segment at grain boundaries. In this case, the flow stress scales inversely with the grain size and the film thickness. Another model to explain higher flow stresses in thin films was proposed by von Blanckenhagen et al. [37]. They simulated the activation of a Frank-Read source as a limiting factor. In their model, dislocation sources are activated multiple times, but the dislocations pile up at interfaces and grain boundaries. Therefore, a backstress is created on the source. The flow stress depends either on film thickness or grain size, whichever is smaller. Experimental investigations on Al and Cu thin films [38-40] validated the different aspects of the proposed models, and come to the conclusion, that depending on the microstructure different models can be applicable. A summary of the proposed models is displayed in Fig. 2.1.

reference	proposed dislocation mechanism under constraints
Chaudhari [30]	minimum energy
Murakami [31]	pinned dislocations bowing out like Orowan mechanism 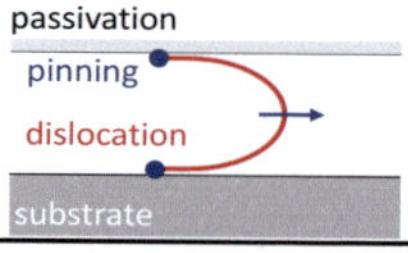
Nix [6], Freund [32]	substrate dislocation (DS) threads thin film and creates misfit DS 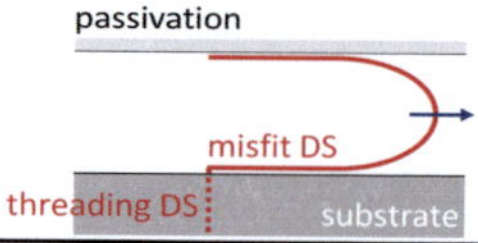
Dörner and Nix [33]	additional Hall-Petch effect due to grain boundaries 
Thompson [36]	grain boundaries (GB) are obstacles to dislocation (DS) motion 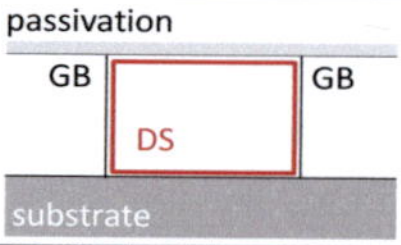
von Blanckenhagen et al. [37]	multiple activation of Frank-Read (FR) sources, pile up at grain boundaries (GB) creates a backstress on the source 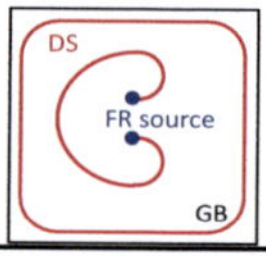

Fig. 2.1: Summary of models explaining the increased flow stress of thin film materials.

In very thin Cu films ($t \sim 100$ nm) Dehm et al. [41], and Balk et al. [42] observed dislocation motion on unexpected slip systems, which they attributed to the inhomogeneous diffusion paths along grain boundaries in columnar grain structures. They argue that this can be interpreted as an indirect evidence of diffusional creep, which was modeled by Gao et al. [43]. This demonstrates that dislocation and grain boundary processes cannot be separated easily and also other microstructural features have to be taken into account. For further reading the review on dislocation plasticity in thin films by Kraft et al. [44] is recommended. With the available computation power, simulations of dislocation behavior gained importance within the last years. A review on simulated dislocation processes and strengths in thin films was compiled by Fertig and Baker [45].

Grain Structure and Interfaces in Thin Films

Grain sizes and boundaries are important microstructural features and have an influence on materials properties. In a classical sense, grain boundaries are regarded as obstacles to dislocation motion. The relation holds over a large range of dimensions even though the mechanisms might change. A proposed mechanism is strengthening by dislocations piling up at the grain boundaries. Hall and Petch [34, 35] found that the strength scales with $d^{-\frac{1}{2}}$. As the grain size is decreased into the nanocrystalline (nc) regime, full dislocations might not be available and other processes determine the strengthening. There is not one single deformation mechanism but a mixture of processes across the length scales, which are still subject of discussion [46, 47]. Possible mechanisms can either be devoted to partial dislocations [48], twinning [49, 50] or diffusion based processes, like grain boundary sliding and grain rotation [51].

Another important aspect to note is that grain boundaries are not stable objects. Grain boundaries can migrate, which is the basic mechanism for grain growth [52]. A driving force for grain boundary migration can be the reduction of an internal energy by adapting the chemical potential, reducing a concentration gradient, or reducing the total grain boundary area. These processes are often thermally activated and based on diffusion processes. In contrast, the grain boundary migration can also be stress induced. One possible process was proposed by Winning et al. [53]. They think that due to a thermal activation grain boundary dislocations can climb and therefore drive the boundary forward. This might be a dominant process at low shear stresses and high homologous temperatures. Another mechanism was introduced by Cahn et al. [54]. They assume that the normal grain boundary motion is coupled to a shear stress. The advancement of the grain boundary is described by an atomic mechanism of coupled motion. This in turn might be a dominant process at high shear stresses and low homologous temperatures.

Interface and surface properties also have a strong influence on the active deformation mechanisms. Experiments demonstrated enhanced properties of thin films with additionally added surface and interface films [55]. Simulations [56] show that the dislocation pile up at hard interface and surface layers. This in turn creates a backstress on the source leading to further strengthening.

Hillock Formation Processes

Hillocks are small bumps at the surface of thin films and usually are observed after deposition or heat treatment. Hillock formation is believed to be caused local compressive stresses [57]. The formation depends on grain boundary motion [58] and diffusion processes along grain boundaries. A passivation layer can change the appearance of the hillock, as diffusion paths along the surface are hindered [59] (Fig. 2.2. (A)) Investigations on the microstructure of hillocks [60] revealed that material can also diffuse along the interface and pushes the thin film upwards (Fig. 2.2. (B)). Hillocks can even be caused by electromigration due to the material transport caused by the electron flow. While the driving force for hillock formation is also an accumulated compressive stress, the formation mechanism is different. Nucci et al. [61] proposed two possible processes for hillock formation. Either the grain rotates first and is then pushed out by grain boundary sliding (Fig. 2.2. (C)), or it is an iterative process where lateral grain boundary motion and grain boundary sliding happens (Fig. 2.2. (D)), which results in a round shape of the hillock.

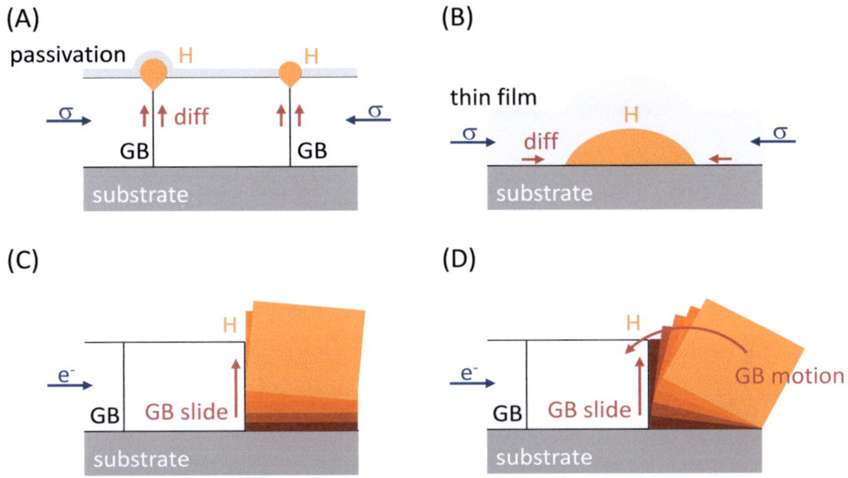

Fig. 2.2: Summary of hillock (H) formation mechanisms. Thermally grown hillocks: in (A) diffusion (diff) along grain boundaries (GB) and influence of passivation layer [59] and (B) diffusion along interface and pushing upwards the thin film [60]. Hillocks grown by electromigration: (C) grain is rotated and pushed out by grain boundary (GB) sliding or (D) alternating process of lateral grain boundary motion and sliding [61].

2.2. Fatigue in Bulk Materials

Fatigue is the phenomenon that a material fails after cyclic loading below the ultimate tensile strength. Fatigue failure was first discovered in the early 19th century [1, 2]. Since then, investigations on reliability of materials reached a high number of publications. In the year 2000 the total number was estimated to be around 100,000 [62]. Thus, there are numerous combinations of causes and effects on the fatigue of materials and only a brief summary of the different aspects of fatigue in bulk materials are pointed out in this chapter. The most popular textbook on this topic is still “Fatigue of Materials” written by Suresh [63] in 1998. It is a comprehensive book on fatigue of bulk materials. A more recent published book is from Schijve [62], which provides a more engineering point of view.

2.2.1. Experimental Options for Fatigue Testing

Bulk fatigue testing can be realized in various experiments. Cyclic loading can either be stress or strain controlled and can be induced mechanically or thermally. The overall loading ratio R can be fully reversed (tension-compression, $R = -1$), pure tensile ($R = 0$), pure compressive ($R = -\infty$), or a combination thereof. The loading condition can be e.g. uniaxial, multiaxial, and experiments can be in tension, compression, bending, torsion, or sliding. Testing frequencies can span from as less than 1 Hz up to the ultrasonic regime of 10^9 Hz. All the previously listed attributes have an influence on the fatigue behavior of materials and the underlying mechanisms and processes. Furthermore, the sample geometry, microstructure, or surface quality, can have positive or negative effects on the fatigue life of a component.

2.2.2. Fatigue Damage Formation in Bulk Materials

Generally, in bulk materials dislocations arrange themselves in wall and cell structures during cyclic loading. By irreversible slip of dislocations on favorable glide planes persistent slip bands (PSB) (see Fig. 2.3) are formed, which can further develop at the surface in extrusions or cracks. Important stages of fatigue damage mechanisms in bulk materials are crack initiation and crack growth [63]. It can be distinguished under which conditions these processes are dominating. This corresponds roughly with the classification of fatigue regimes. Fatigue failure within the first 10^4 cycles is called the low cycle fatigue (LCF), where the sample is plastically strained and crack growth is dominating. Above 10^4 cycles the sample is strained elastically. Failure between 10^5 and 10^8 cycles is called high cycle fatigue (HCF), which is dominated by crack initiation. The regime above 10^8 cycles is called very high cycle fatigue (VHCF) and dominating processes are proposed to be crack initiation by strain localization [64]. Furthermore, at these high cycle numbers the microstructure and defects gain more importance for the damage mechanisms as discussed in [65]. Fcc materials seem to have no endurance limit. As fatigue experiments evolved it was found that with increasing cycle numbers the fatigue damage moves on [66] and a multi-stage fatigue life was proposed [67].

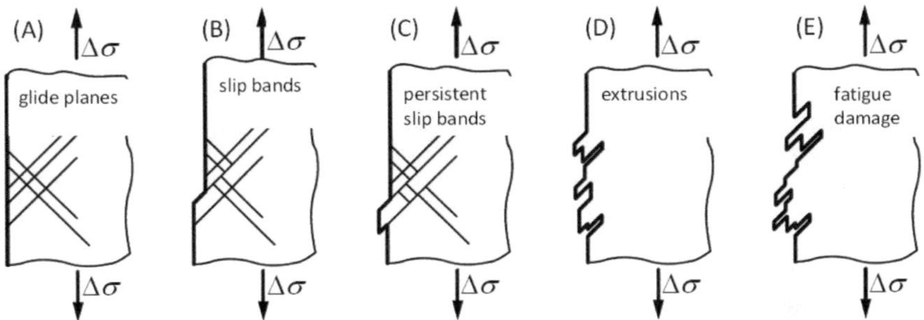

Fig. 2.3: Irreversible slip of dislocations on favorable glide planes creates slip bands. With increasing cycle numbers from (A) to (E) they become persistent and extrusions are formed at the surface. Taken from [68].

Wöhler [2] established to plot the lifetime data in so called stress amplitude versus number of cycle (S-N) curves. Empirical laws describing the trend for the strain ranges in the LCF and the HCF regime were introduced by Coffin and Manson (Eq. 2.1) [3, 4], and Basquin (Eq. 2.2) [5], respectively:

$$\frac{\Delta\varepsilon_{pl}}{2} = \varepsilon'_f \left(2N_f\right)^c \qquad \textit{Eq. 2.1}$$

and

$$\frac{\Delta\varepsilon_{el}}{2} = \frac{\sigma'_f}{E} \left(2N_f\right)^b \qquad \textit{Eq. 2.2}$$

With $\Delta\varepsilon_{\mathrm{pl}}$ and $\Delta\varepsilon_{\mathrm{el}}$ being the plastic and elastic strain range, ε_{f}‘ is the fatigue ductility coefficient, $2N_{\mathrm{f}}$ are the load reversals to failure, c is the fatigue ductility exponent, σ_{f}‘ is the fatigue strength, and b is the fatigue sensitivity exponent.

Gaining deeper insights into fatigue damage, more specified model conceptions were proposed for example by Essmann et al. [69]. They suggested that vacancies are produced by the annihilation of dislocations, which in turn leads to an elongation of PSBs and finally an extrusion. Going from single crystalline samples to polycrystalline samples, it was found that slip orientation and grain boundaries have an influence on the dislocation motion [70]. The orientation of the grain sets a threshold for dislocation slip, and the grain boundaries cause the dislocations to pile up. Thus, intergranular fatigue cracking is more pronounced due to interactions of PSBs and grain boundaries. Reducing the grain size even more into the ultra fine-grained (ufg) [71] or even the nanocrystalline (nc) regime [72], it was found that fatigue life is reduced in the LCF regime, due to the loss in ductility, but it is increased in the HCF regime, due to strengthening effect of the grain boundaries. More complicated mechanisms dependent on strain rate-dependent diffusion processes along grain boundaries have to be taken into account.

2.3. Fatigue in Thin Films

In the following chapter a literature review on fatigue in thin films is conducted. First, several testing techniques will be presented. Furthermore, results from previous work on freestanding thin films as well as substrate bonded thin films are summarized. The main difference between freestanding thin films and thin films on a substrate is that freestanding thin films fail in a fatal manner if cracking occurs. Thin films bonded to the substrate can display cracking but, due to the support, would not fail fatally. Another main difference is the confinement by the substrate itself as deformation processes might be hindered. An overview on fatigue of small scale metal materials is given by Zhang and Wang [73].

2.3.1. Experimental Techniques for Thin Film Fatigue Testing

Over the past years, various testing techniques for thin film fatigue testing have been developed and are presented in their chronological order. Read et al. [74] developed a uniaxial tension-tension setup for testing freestanding thin films with piezoelectric stacks as actuators. The displacement is measured and controlled by non contact eddy-current sensors. To test thin films on a stiff substrate Schwaiger et al. [75] proposed cyclic bending of a micro cantilever by using a nanoindenter. The thin films were subjected to fatigue at the highest strains caused by the bending moment. The test was monitored by changes in the stiffness of the cantilever. To study thermal fatigue of thin Cu lines, Mönig et al. [76] introduced a setup, where the Cu lines are thermally cycled by an applied alternating current. Eve et al. [77] developed an LCF setup for thin films on polymer substrates, which induces a biaxial stress state in the film equivalent to the one created by thermal cycling. They used a ring-on-ring test and included an *in situ* optical failure detection system. For self supported Cu/Nb multilayer samples Wang et al. [78] developed a resonant vibrating setup. They actuated their cantilevers with a bimorph piezoelectric plate-like actuator and measured the deflection by a fiber optic measurement system. To reach extreme high cycle numbers up to 10^{14} Eberl et al. [79] used surface acoustic waves (SAW) with frequencies in the MHz regime to fatigue thin Al lines. Wang et al. [80] introduced another technique. They deposited a thin Cu film on a dog bone shaped polyimide substrate. Due to the compliance of the substrate, the thin film is deformed in tension and compression by loading and unloading the whole sample. The stress-strain curves were measured by *in situ* X-Ray Diffraction (XRD). As flexible electronics became popular, fatigue testing techniques were developed for this application [81]. They used similar thin film samples on a polymer substrate and measured the electrical resistance of the thin film during cyclic loading.

Table 2.1: Summary of thin film fatigue testing techniques and their failure criteria.

reference	substrate	mode	R	f (Hz)	failure criterion	comment
Read et al.[74]	free	tensile	0.1	0.07	cannot reach loading amplitude	
Schwaiger et al. [75]	Si	bending	-	45	stiffness decrease	nanoindenter
Mönig et al. [76]	Si	thermal	-	100	electrical open	in situ SEM
Eve et al. [77]	polymer	biaxial	-	0.2	stray light	
Wang et al. [78]	free	bending	-1	800	fracture	resonance
Eberl et al. [79]	Si	SAW	-	900 M	frequency shift	
Wang et al. [80]	polymer	tensile	-1	100	saturation of extrusion density	in situ XRD
Sun et al. [81]	polymer	tensile	0.05	5	resistance	

All the presented testing systems are summarized in Table 2.1. They all have advantages and disadvantages with respect to the sample preparation, the possible number of cycles, the applied cyclic loading and the definition of a failure criterion. Nevertheless, no particular testing method accomplished to become a standard testing technique.

2.3.2. Fatigue in Freestanding Thin Films

Experimental results of freestanding thin films will be presented in this chapter. For different kinds of freestanding thin films with thicknesses above 10 μm, it was found that fatigue cracks can still originate from extrusions as observed in bulk material [73]. In contrast, an absence of clear slip steps for thinner films with thicknesses less than 10 μm was observed. This is attributed to a size effect, as in a small volume dislocations can only move individually [73].

Testing thicker freestanding films, Yang et al. [82] investigated the fatigue behavior of nanocrystalline Ni foils with a thickness of 70 μm up to 10^7 cycles. They observed extrusion formation and crack initiation at surface defects, after significant changes in the microstructure, mostly local grain coarsening. Wang et al. [83] investigated the fatigue properties of nanoscale Cu/Nb multilayers with a total thickness of 40 μm and an individual layer thickness of 40 nm up to 10^9 cycles. The samples fractured at the position of highest stresses but did not show any kind of extrusion formation. It was speculated that this might be due to constraints due to the individual thin layers. Furthermore, the Cu/Nb multilayers displayed a fatigue limit at around 1/3 of its ultimate tensile strength. Read et al. [7, 74] tested freestanding Al and Cu thin films with a thickness of 1 μm and grain size 1 μm in tension-tension up to 10^5 cycles. They observed that the Cu thin films had longer lifetimes than the Al thin films. Possible failure mechanisms in microstructure were related to plastic ratcheting (progressive accumulation of deformation) in the LCF regime and slip lines in the HCF regime. Bae et al. [84] tested freestanding 1 μm Al thin films up 10^7 cycles under closed loop load control and found a roughening of the Al surface. All these experiments support the fact that an increasing surface to volume ratio and the microstructure have an effect on the fatigue behavior.

2.3.3. Fatigue in Thin Films on a Substrate

First, experimental results of thin films deposited on a compliant substrate are presented. Al, Au or Cu thin films on polyimide were tested by several authors [80, 81, 85-91]. Especially the investigations presented in [80, 88, 91] focused on the influence of film thickness or grain size on the fatigue life and damage formation. Generally, it was found that with decreasing film thickness or grain size the fatigue lifetime increases. Whatever has the smaller dimension has the stronger influence on the fatigue damage processes. Detailed microstructural observations of Cu thin films fatigued in the LCF regime revealed three possible damage formations types for thin films with large grains [91]. Thin films of a thickness above 3 μm displayed a bulk-like behavior as the surface had a high amount of large extrusions. TEM investigations obtained dislocation structures like walls and cells similar to bulk material. Thin films in the range of 1 μm were found to be in a transition region as the number and size of the extrusions decreased and tangled dislocation structures were observed. Small scale behavior was examined for even thinner films below 1 μm. In this case cracking along grain boundaries was more pronounced than extrusion formation. In the TEM micrograph only single dislocations could be found. If the grain size is smaller than the film thickness as observed in [88, 91], damage was dominated by roughening of the surface followed by cracking along grain boundaries. Another feature observed in Cu thin films [80, 90] was that extrusion formation came along with pore formation at the interface.

Secondly, thin films were tested on stiff Si substrates. The fatigue behavior of Ag, Al or Cu thin films on Si substrate were investigated for mechanical [75, 79, 92, 93] and thermal [76, 94-96] loading. Similarly to thin films on compliant substrates the mechanically loaded Ag thin films displayed extrusion and void formation [75, 93]. Furthermore, it was found that (100) oriented grains are more pronounced to extrusion formation due to smaller flow stresses compared to

(111) oriented grains [93]. Additionally, a size effect in these Ag thin films depending on the film thickness was found. Somewhat different is the situation for the Al lines tested to ultra high cycle numbers [79, 92]. A mixture of extrusion formation, grain growth, hillock formation and grooving was found. These damage formations could be attributed to the very high frequencies and different activated processes like dislocation climbing [97]. Even after 10^{14} cycles no endurance limit was found. In the case of the thermally cycled Cu thin films [94, 95], severe surface damage was examined. Extrusion formation and grain growth was observed in individual grains depending on their orientation.

2.3.4. Damage Models in Thin Films

The experimental investigations on the fatigue of thin films showed different damage structures compared to bulk materials. Taking the characteristic of thin films into account it was attempted to propose a model for damage formation in thin films. A first effort was done by Schwaiger et al. [75] by combining the Nix model for dislocations in thin films [6, 32] with the proposed fatigue damage formation by Essmann et al. [69]. If dislocations have opposite signs and are close enough to each other, they annihilate. By accumulation of this process, the void formation at the substrate interface could be understood. The model was refined by Schwaiger et al. [90] and was enhanced by adding diffusion processes of vacancies.

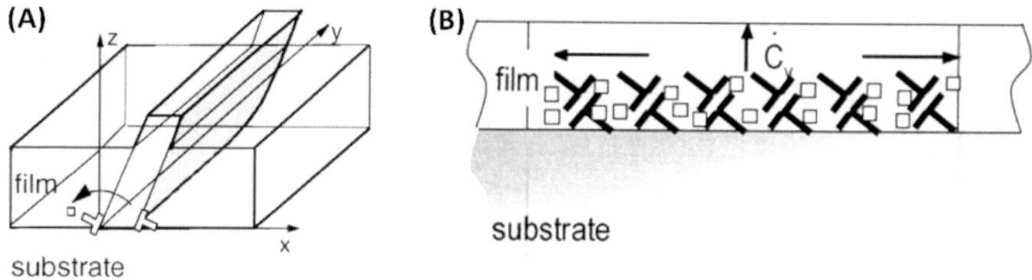

Fig. 2.4: Proposed mechanism for pore formation in cyclically deformed Cu thin films on a substrate. (A) Vacancy formation due to dislocation annihilation [75] and (B) influence of vacancy diffusion [90].

Zhang et al. [91] discussed a size effect by the ability of forming dislocation structures with respect to several influences on dislocation nucleation and motion. First, the proximity of the free surface plays a role, as image forces are high near the surface and tend to draw dislocations out of the thin film. Secondly, they proposed that there is a minimum length scale for dislocation self-organization at around 1 µm. And thirdly, the already mentioned constraints on dislocation nucleation and motion observed in thin films apply.

2.4. Summary and Motivation

The literature review in the previous sections demonstrated that there are several mechanisms occurring during monotonic as well as cyclic loading. The mechanisms are based on dislocation, grain boundary or diffusion processes. Furthermore, these processes often cannot be separated, but instead happen simultaneously and strongly depend on the microstructure of the material.

So far research was focused on mechanical properties of thin film materials under monotonic loading. In recent years, additionally the investigations on fatigue properties of thin films have evolved. Predominantly, thin films were tested in the LCF regime obtaining damage structures and mechanisms with the focus on size effects. Unfortunately, as working frequencies are higher than in macro devices, little research was conducted in the HCF regime. Thus, this is mandatory as small scale devices in microelectronics and MEMS have to withstand high cycle numbers. Therefore, there is a need to test and understand fatigue behavior of thin films in the HCF and VHCF regime in a reasonable amount of time. For this purpose, a new kind of testing method is needed, which was accomplished in this thesis.

The aim of this thesis is to develop a novel high-throughput bending setup for thin film fatigue testing. With the idea of inducing a strain gradient along a cantilever in a thin film, the damage evolution can be monitored depending on the cycle number. Thus, it will be possible to obtain one lifetime diagram with only one sample. Furthermore, the experiments should cover the HCF and VHCF regime. The two materials Al and Cu are investigated as leadoff materials. Cu is the typical model material, and much research was done on fatigue of Cu thin films. In comparison, the fatigue behavior of Al thin films has not been investigated so well. With these two materials, several influences can be investigated. Al has a passivated surface and a high homologous temperature at room temperature compared to Cu with a free surface and a low homologous temperature. Furthermore, the effect of an adhesion layer is studied by additionally depositing a Ti thin film underneath the Cu. Moreover, this work focuses on influences of the microstructure. Fine-grained and coarse-grained thin films for both materials are tested. The novel experimental method allows deriving a phenomenological model which can describe the damage growth induced by fatigue. This can be used to establish a lifetime diagram depending on the fraction of damage at a given strain amplitude after a specific number of cycles.

3. Development of a Resonant Bending Setup for High-Throughput Thin Film Fatigue

Usually fatigue testing is tedious work, because one needs a lot of samples for different stress/strain amplitudes and a lot of time if high cycle numbers are to be reached. Thinking in new directions will open up new possibilities to test materials in an efficient way. In this chapter a novel fatigue test setup for thin films on Si substrate will be presented. At first, the working principle will be introduced, with basics of the analytical beam bending theory. Here, Finite Element (FE) Simulation plays an important role for calculating strains and stresses in the thin film, and to develop a strain calculation routine. In the end of this chapter the implementation of the setup is presented, concluding with an overall discussion of problems encountered.

3.1. Basic Idea of a Cantilever Bending Fatigue Setup

The analytical beam bending theory is the basis for the novel cantilever bending setup for the fatigue testing of thin films on a stiff substrate. All the details described below are in principle transferable to other size scales and materials. Nevertheless, the focus of this study is small scale testing of cantilevers with an on top deposited thin film.

3.1.1. Derivation of Surface Strains Induced by Bending Deformation

Starting from a stiff cantilever fixed at one end and displaced in the z-direction by a normal point load P_z at position $x = L$ (Fig. 3.1 (A)). L is the length of the cantilever and also defines the distance of the fixed end to the free end. Assuming pure bending, isotropic elastic behavior, and small displacements, the special case of Bernoulli [98] can be assumed. According to this analytical beam bending theory, the resulting bending moments $M_z(x)$ depend on load P_z, the position x, and the length L of the cantilever. That is:

$$M_z(x) = P_z(x - L) \qquad \textit{Eq. 3.1}$$

The bending moment causes a surface stress $\sigma_z^{surf}(x)$ at position x which can be calculated by

$$\sigma_z^{surf}(x) = \frac{M_z(x) t_{sub}}{2I} \qquad \textit{Eq. 3.2}$$

where t_{sub} is the cantilever substrate thickness and I is the second moment of inertia at position x. A constant rectangular cross section is assumed throughout the whole cantilever length, whereas I can be correlated to the width w and the cantilever thickness t_{sub} in the following equation:

$$I = \frac{w t_{sub}^3}{12} \qquad \textit{Eq. 3.3}$$

Therefore, applying a load at the free end of a cantilever results in a stress gradient in the x-direction at the surface of the cantilever. This stress $\sigma_z^{surf}(x)$ can be calculated into a strain $\varepsilon_z^{surf}(x)$ by using Hookes law:

$$\varepsilon_z^{surf}(x) = \frac{\sigma_z^{surf}(x)}{\hat{E}} \qquad \textit{Eq. 3.4}$$

where $\hat{E} = E$ in the case of a slim cantilever (this case), or $\hat{E} = E/(1-\nu^2)$ in the case of a broad cantilever. E is the Young's Modulus of the material and ν the Poisson's ratio. The situation of the resulting strain gradient on the sample surface along the cantilever is shown in Fig. 3.1 (B). An important aspect to keep in mind is that the strain amplitude on the surface $\varepsilon_z^{surf}(x)$ correlates linearly with the position x along the cantilever.

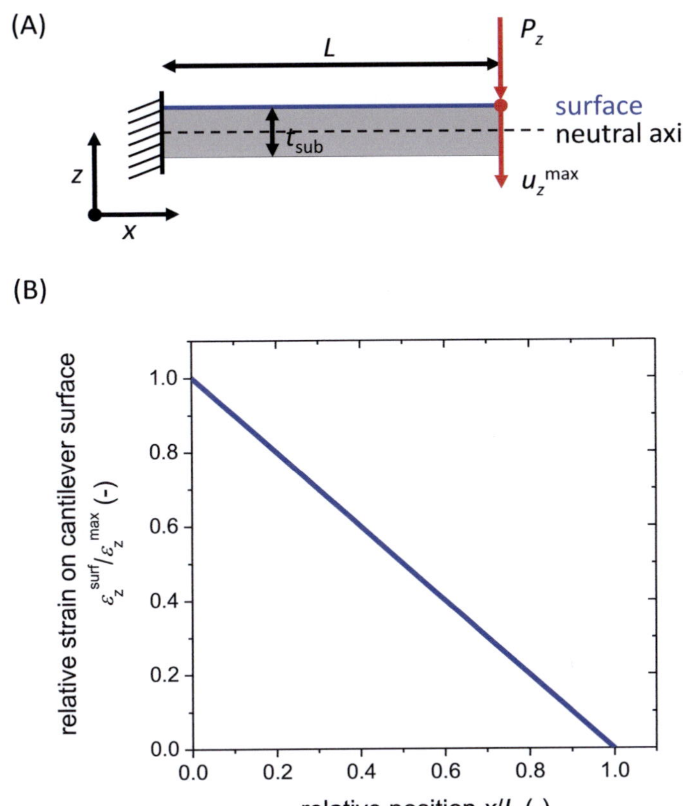

Fig. 3.1: (A) Analytical cantilever bending theory: Cantilever fixed at one end and displaced by u_z^{max} at the free end at $x = L$. (B) Plotting the relative strain amplitude on the cantilever surface over the relative position results in a linear strain distribution along the cantilever.

Instead of applying a point load P_z, it is equal to displace the cantilever by u_z^{max} in z-direction at position $x = L$:

$$u_z^{max} = \frac{P_z x^2 (3L - x)}{6\hat{E}I} \qquad \textit{Eq. 3.5}$$

Combining equations Eq. 3.1-Eq. 3.5 results in Eq. 3.6 for the case of cyclic loading, where the strain amplitude acting in the thin film equals the surface strain induced by bending ($\varepsilon_a(x) = \varepsilon_z^{surf}(x)$) for every position x along the cantilever. This can be calculated from:

$$\varepsilon_a(x) = \frac{3u_z^{max} t_{sub}}{2L^3}(x - L) \qquad \textit{Eq. 3.6}$$

The length L of the cantilever is constant, but the cantilever thickness t_{sub} can vary due to the fabrication process. The displacement amplitude u_z^{max} is measured during the experiment.

If the resonance frequency f_{res} of the cantilever is known, it is possible to excite the cantilever without touching it. The resonant frequency f_{res} of a cantilever depends on the mass of the cantilever m_0 and the weight m_{att} of the attached mass, as displayed in the following equation:

$$f_{res} = \frac{1}{2\pi} \cdot \frac{3.52}{L^2} \cdot \sqrt{\frac{EIL}{m_0}} \cdot \sqrt{\frac{m_0}{m_0 + m_{att}}} \qquad \textit{Eq. 3.7}$$

3.1.2. From Theory to Implementation

With all this basic knowledge mentioned in the previous paragraph an intriguing principle is given for testing cantilever shaped samples in a cyclical bending manner. Regarding a deposited thin film on top of a stiff cantilever, and assuming the thin film approximation, the surface strain gradient along the cantilever is fully transferred into the thin film. Due to this strain gradient, fatigue testing with strain amplitudes from 0 to ε_a^{max} can be conducted with only one sample, which gives the opportunity for high-throughput testing. If the cantilever is cycled at its resonant frequency, the first fatigue damage will occur at the site of the highest strain, namely near the shoulder. If the cantilever substrate can sustain a higher amplitude than the thin film, only the thin film is subjected to fatigue. Following the methodology explained above, a fatigue damage front will move along the strain gradient of the cantilever. Finally, the position of the damage front can be correlated with the acting strain amplitude and the number of cycles and can be used to determine a lifetime diagram (S-N curve), which will be explained in chapter 3.3.3.

As mentioned before, all the information given is transferrable to other size scales and materials. If something other than a coating is subjected to cycling, such as a bulk sample, then gradients throughout the thickness also have to be taken into account, which is not subject to this study. In the following chapters micro Si cantilevers with deposited thin films are the focus of interest.

3.2. Finite Element Simulation of Cantilever Bending

The micro samples used in this work are Si cantilevers coated on top with magnetron sputter deposited 1 μm thin metal films of Al or Cu. The presented cantilever geometry will serve as an example of how to develop a model to calculate the strains acting on the thin film material. The Finite Element (FE) simulation plays an important role for the calculation of the strain and corresponding stress amplitudes acting in the deposited thin film, as these magnitudes cannot be measured directly with the setup. To get reasonable data, the cantilevers were modeled with different element types, analysis methods and various substrate thicknesses to study the influence on strain and resonant frequency.

3.2.1. Finite Element Simulation Details

For the basic FE simulation a simplified geometry is chosen, which has been adapted with different parameters according to the simulation problem. The shape and geometry of the cantilever is shown in Fig. 3.2. The cantilever is in general 10 mm long, 1 mm wide and 200 μm thick. The support of the cantilever, which acts as the fixed end, has a thickness of 500 μm and the shoulder forms an angle of 45°. These are the basic dimensions of the cantilever except when stated otherwise. At the end of the cantilever two 1 mm^3 cubic masses are attached. By varying the densities of these unit masses, different weights are attached to the cantilever. As described later, these masses in the real experiment are Pb spheres, with different weights, ranging from 5 - 12 mg. The defined boundary conditions are zero degrees of freedom (DOF) at the very left end of the support and the flanks have mirror planes to reproduce periodicity.

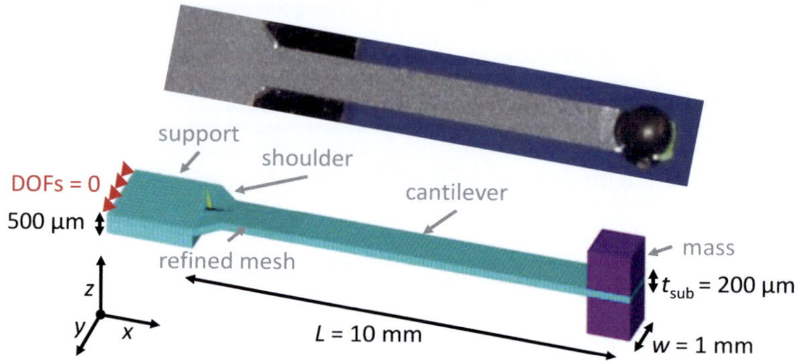

Fig. 3.2: FE model of the Si cantilever with basic dimensions of 10 x 1 x 0.2 mm and an attached cubic unit mass with variable densities. The mesh is refined near the fixed end. For comparison an image of a real cantilever is shown at the top.

Si was chosen as substrate material, as it is stiff, widely used in micro technology, and the etching processes are well developed. The material parameters used for simulation for Si are Young's Modulus $E_{[100]} = 131$ GPa, Poisson's ratio $\nu = 0.28$, and density $\rho = 2.33$ g/cm³. Only the uncoated Si cantilever is simulated without the deposited thin film, as the thin film approximation (about 1 µm thin film to 200 µm substrate) states that the strains and stresses calculated at the top surface of the cantilever are fully transferred into the thin film without a thickness gradient. Also, the thin film is so thin that it does not alter the properties of the cantilever in a significant manner. Changing the thickness of the cantilever will have an influence on the resulting strains, which will be shown later.

ANSYS (academic 12.1, ANSYS Inc., USA) was used to perform the FE simulations. The cantilever was simulated in a 3 D model with eight-node elements (SOLID 45) with three degrees of freedom in plain strain. The simulations were done for purely elastic and isotropic material properties and damping was neglected. The analysis type was either chosen to be linear 'static' or 'modal', for comparison. In the 'static' analysis a predescribed displacement in z-direction is applied to the center node at the end of the cantilever. The 'modal' analysis determines the vibration characteristics like resonant frequencies and mode shapes of a structure. As the region near the fixed end is of specific interest, the mesh was refined by a ratio of 1 to 3 near the shoulder.

3.2.2. Evaluation of Simulation Parameters via 'Modal' Analysis

To establish a strain gradient calculation routine, a parameter study was conducted. This was accomplished by using the 'modal' analysis. First, the element size was varied to optimize the simulation results while keeping calculation time at a reasonable value. Additionally, the potential change, related to influences from the attached mass or substrate thickness, is monitored in the first three resonant modes of excitation.

The first three resonant modes of the cantilever are shown in Fig. 3.3. The first mode shows an out-of-plane bending of the cantilever, the second mode is an in-plane bending and the third mode is torsion. The three modes occur at different resonant frequencies. In the beginning all three modes are investigated for changes in their principle behavior.

With the 'modal' analysis in ANSYS the resonant frequency can be calculated (Block Lanczos solver). In this case a mass of 12 mg (2 x 6 mg) was attached to the free end of the cantilever. The element size was varied between 50 - 2000 µm for the first three resonant modes. In the diagram in Fig. 3.4 on the left axis the resonant frequencies of the first three modes are plotted versus the element size. On the right side, there is the corresponding calculation time plotted versus element size. With decreasing element size, the three resonant frequencies decrease and saturate for the smallest element sizes. Further downscaling of the element size does not change the resonant frequency significantly. The overall trend of the resonant frequencies for the different modes is similar, whereas the out-of-plane bending lies by the smallest frequency. There is no crossover in the frequencies of the three modes (be aware of the different scaling of the axes). Therefore, it is safe to only investigate the out-of-plane bending in further calculations. With decreasing element size the calculation time increases exponentially. For the following calculations an element size of 100 µm (indicated by a grey line) was chosen, which is a good compromise for a satisfactory solution and a reasonable calculation time.

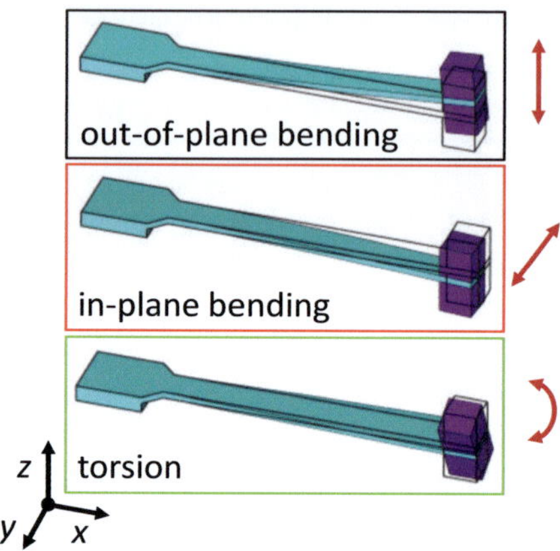

Fig. 3.3: First three resonant modes of the Si cantilever. The first mode is out-of-plane bending, the second mode is in-plane bending, and the third mode is torsion.

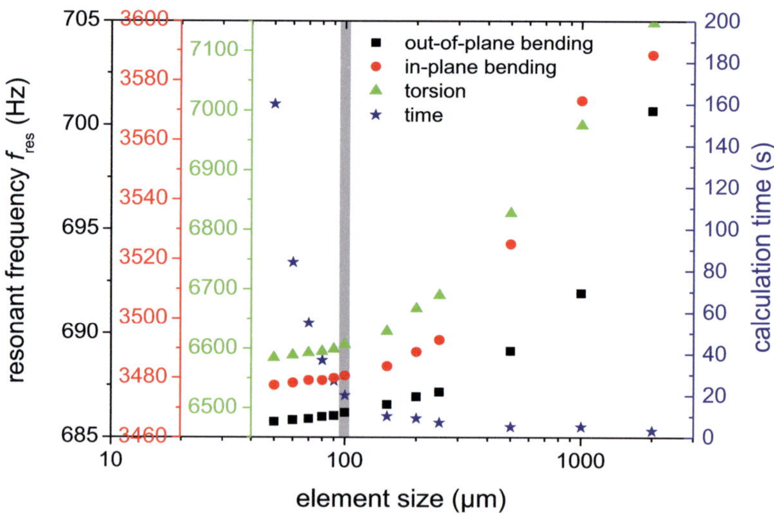

Fig. 3.4: Influence of element size on the resonant frequency of the first three modes and the overall calculation time. On the left axes, there are the three resonant frequencies with different scales for each mode and on the right side is the corresponding calculation time. The resonant frequencies decrease with decreasing element size but the calculation time is increasing. An element size of 100 µm is regarded to be a good compromise indicated by a grey line.

3.2.3. Comparison of 'Static' and 'Modal' Analysis

The next step for establishing a calculation routine for the strain gradient in the thin film is to compare the analytic solution of a one side fixed cantilever with a 'static' and 'modal' FE analysis. Fig. 3.5 (A) shows a schematic comparison of the analytic solution, 'static' and 'modal' analysis. For the analytic solution Eq. 3.5 and Eq. 3.6 were used. The 'static' analysis was performed by displacing the free end in the center of the cantilever by $u_z^{max} = 1000$ µm. In the 'modal' analysis the program calculates the resonant frequencies and the resulting stresses on the surface. To compare it with the 'static' analysis the results for stress and strain of the 'modal' analysis are normalized to the same displacement $u_z^{max} = 1000$ µm at the free end. In both FE analysis cases a path in the surface center of the cantilever was defined, to obtain the calculated stress and strain data. The attached mass was chosen to be 12 mg.

In Fig. 3.5 (B) the displacement amplitude u_z of the different cantilevers is plotted versus the position along the cantilever. All curves lay over each other and are in good agreement. The situation for the strain amplitude ε_a at the top surface of the cantilevers in (C) is a little different. As expected, the strain amplitude on the surface decreases linearly from the shoulder towards the free end of the cantilever. The trend is similar for all the three cases with small differences in slope and maximum value, simply because the analytic solution is an ideal case where there is no mechanical transition from the support to the fixed end. Also the mass at the free end of the simulated cantilevers has a small influence on the strain gradient curve. In conclusion, the simulations provide reasonable results compared to the analytic solution. For the maximum strain amplitude ε_a^{max} there is a relative discrepancy of roughly 3 %.

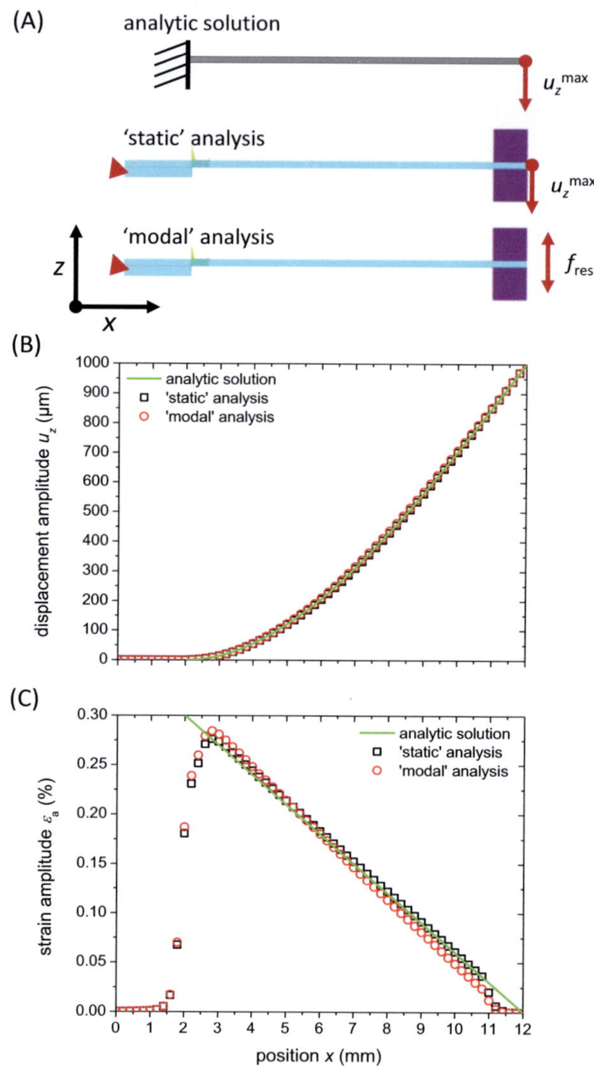

Fig. 3.5: Comparison of analytic solution, 'static' and 'modal' analysis of a cantilever fixed on one side (at x = 2 mm) and displaced at the free end. (A) Schematic side view of the cantilevers. (B) Displacement amplitude u_z versus the position x along the cantilever for u_z^{max} = 1000 µm. (C) Strain amplitude ε_a on the cantilever surface along the positions x of the cantilever.

3.2.4. Influences of Mass and Cantilever Thickness

The next step for a reliable FE analysis is to investigate the influence of the attached mass on the strain amplitude at the surface of the cantilever, which is shown in Fig. 3.6. The total weight of the two masses was varied between 1 mg and 22 mg. For the 'static' and 'modal' analysis a maximum displacement amplitude u_z^{max} was chosen to be 1000 µm. The 'static' analysis of the cantilever displacement shows a constant maximum strain amplitude $\varepsilon_{\mathrm{a}}^{\mathrm{max}}$ independent of the attached weight of the mass, since the displacement of the free end u_z^{max} is constant. In contrast to that, the maximum strain amplitude $\varepsilon_{\mathrm{a}}^{\mathrm{max}}$ increases for lighter masses in the 'modal' analysis. This is due to a slightly different stress distribution in the cantilever and higher resonant frequencies, and the normalization of the displacement to u_z^{max}. To minimize the influence of the mass, only a total weight between 9 - 22 mg will be used in the experiment. As the maximum strain amplitude is constant in that region, the mass is not subjected to further simulations, but an error of 3 % has to be taken into account, which is also indicated by the error bars of the displacement analysis.

Furthermore, the influence of the attached mass on the resonant frequency of the first three modes was calculated, which is shown in Fig. 3.7. With decreasing weight of the mass the resonant frequency increases for all three resonant modes and there is no crossover between the modes. Therefore, it is appropriate to simulate only the out-of-plane bending in further calculations, as that is the desired excitation mode. The bump in the green curve of the torsion mode for small masses can be explained by a possible mode transition, due to the fact that the weight of the attached mass is in the range of the weight of the cantilever itself ($m_0 \sim 4$ mg). As these small weights are not used in the experiment and the frequency range is also not considered, this effect can be neglected.

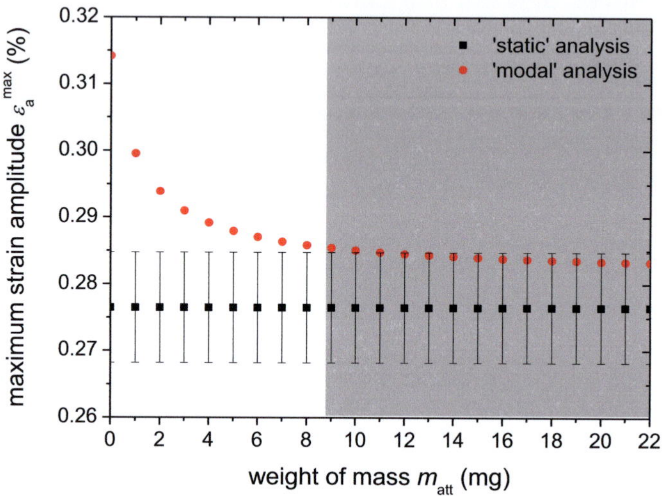

Fig. 3.6: Maximum strain amplitude ε_a^{max} near the shoulder at the surface of the cantilever depending on the weight of the attached mass at the free end of the cantilever for the 'static' and 'modal' analysis for a maximum displacement amplitude u_z^{max} of 1000 µm.

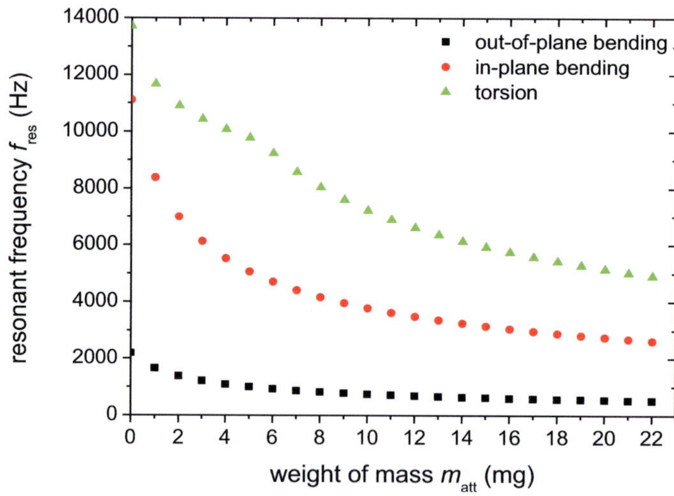

Fig. 3.7: Resonant frequency versus the weight of the attached mass for the first three modes in the 'modal' FE analysis.

The real cantilever thickness is variable and depends on the fabrication process of the Si substrate. The overall stiffness of a thinner cantilever is smaller than for a thicker one and therefore changes the resonant frequency, which is displayed in Fig. 3.8. Additionally, the dependence of the attached mass can be seen. The resonant frequency increases with increasing cantilever thickness t_{sub} and decreasing weight of the mass m_{att}. With this diagram the resonant frequency in the real experiment can be estimated.

As the cantilever thickness t_{sub} is a crucial variable and is linearly correlated to the strain amplitude ε_{a} (see Eq. 3.6) the influence on the maximum strain amplitude $\varepsilon_{\mathrm{a}}^{\mathrm{max}}$ is simulated. The diagram in Fig. 3.9 illustrates the dependence of the maximum strain amplitude $\varepsilon_{\mathrm{a}}^{\mathrm{max}}$ for substrate thicknesses $t_{\mathrm{sub}} = 150 - 200$ µm with respect to variable displacement amplitudes $u_z^{\mathrm{max}} = 0 - 1000$ µm. The maximum strain amplitude $\varepsilon_{\mathrm{a}}^{\mathrm{max}}$ correlates linearly with the displacement amplitude u_z^{max} as expected from Eq. 3.6. Furthermore, the maximum strain amplitude $\varepsilon_{\mathrm{a}}^{\mathrm{max}}$ increases with increasing substrate thickness t_{sub} It is now desirable to have an equation to calculate the maximum strain amplitude $\varepsilon_{\mathrm{a}}^{\mathrm{max}}$ by entering the actual substrate thickness t_{sub} and the reached maximum displacement amplitude u_z^{max}. The shown cantilever bending is no ideal case, the analytical solution overestimates the maximum strain amplitude. Thus, a calculation routine was established based on FE simulations. How this was done will be explained in the following section.

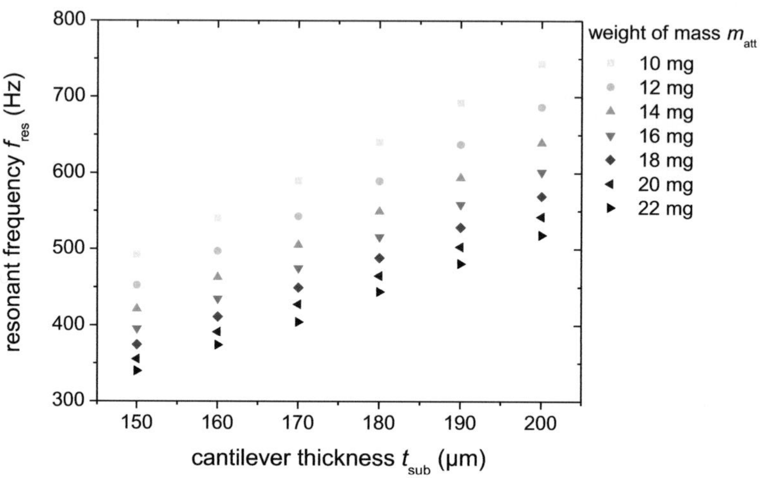

Fig. 3.8: Resonance frequency f_{res} versus cantilever thickness t_{sub} with variable masses attached at the free end.

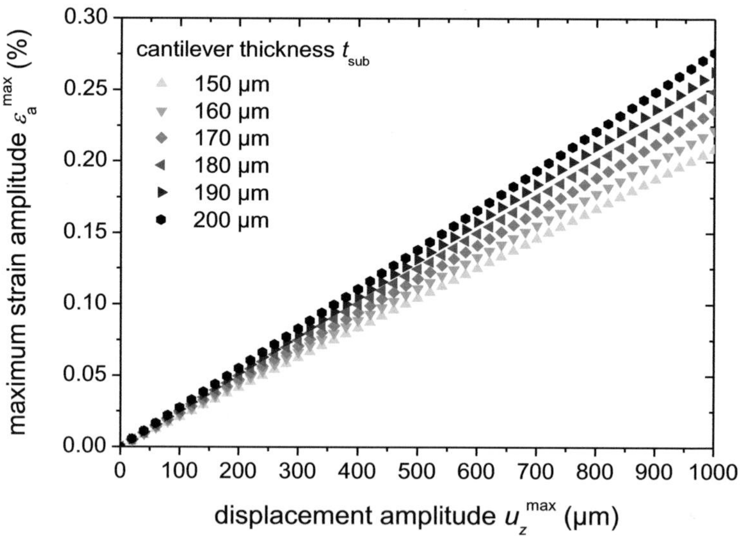

Fig. 3.9: Maximum strain amplitude ε_a^{max} at the surface of the cantilever shoulder versus different maximum displacement amplitudes u_z^{max} for different cantilever thicknesses t_{sub}.

3.2.5. Calculation of Maximum Strain Amplitude

The aim of this chapter is to deduce an equation to calculate the maximum strain amplitude ε_a^{max} with respect to a specific cantilever thickness t_{sub} and maximum displacement amplitude u_z^{max}. These two variables depend on the sample processing beforehand and the excitation during testing. Therefore, there is the need for calculating the maximum strain amplitude ε_a^{max} for all possible parameter combinations.

'Static' and 'modal' FE analysis were conducted for different cantilever thicknesses with a displacement amplitude u_z^{max} of 1000 μm and an attached mass m_{att} of 12 mg. The calculated maximum strain amplitude ε_a^{max} at the surface of the cantilever near the shoulder is plotted versus the thickness t_{sub} (Fig. 3.10). The cantilever thickness was chosen to be in the possible range of the actual samples (t_{sub} = 150 – 200 μm). In this range the simulated data was fitted linearly, with the resulting fitting equations displayed in the diagram. The discrepancy between the 'static' and the 'modal' solution is again about 3 % for these thickness values. It is possible that the difference increases for thicker substrates because of the change in ratio of mass of the cantilever m_0 and attached mass m_{att}. For further calculations of the maximum strain amplitude ε_a^{max}, the more conservative case ('static' simulation) has been used. In combination with the fact that the strain amplitude ε_a correlates linearly with the maximum displacement amplitude u_z^{max}, the final result for the calculation of the maximum strain amplitude ε_a^{max} is given by

$$\varepsilon_a^{max} = \left(A + B \cdot t_{sub}\right) \cdot u_z^{max} \qquad \textit{Eq. 3.8}$$

with $A = 4.48 \times 10^{-6}$ %/μm and $B = 1.36 \times 10^{-6}$ %/μm^2, where t_{sub} and u_z^{max} have to be inserted in μm to get the strain amplitude in %. The calculated strain amplitude curve in Fig. 3.5 (C) can be normalized to the specific maximum strain amplitude ε_a^{max} for each cantilever. With this methodology a fairly simple calculation gives all the information needed for calculating the strain gradient in a thin film along the Si substrate cantilever.

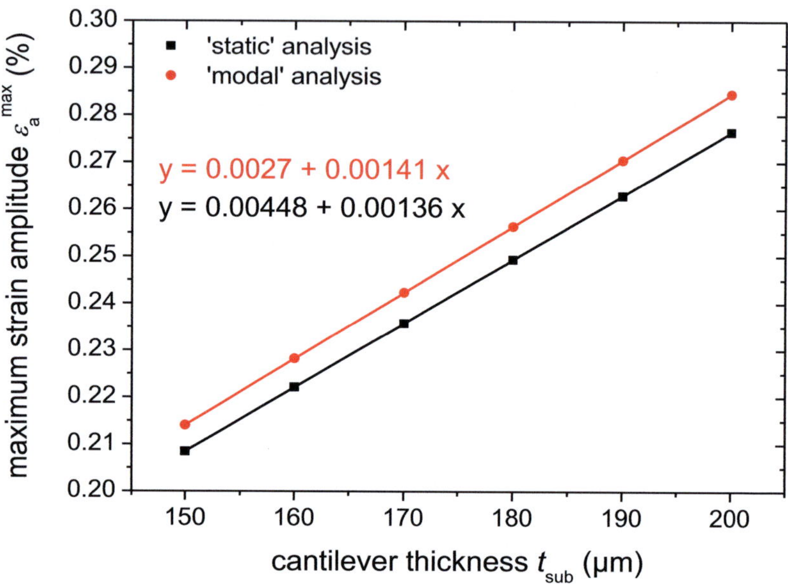

Fig. 3.10: Linear fit of maximum strain amplitude ε_a^{max} versus cantilever thickness t_{sub} for the 'static' and 'modal' case at a maximum displacement u_z^{max} = 1000 μm and a cantilever length L = 10 mm.

3.3. Implementation of the Bending Fatigue Setup

Based on the theoretical considerations of the previous sections, a setup was designed to test the cantilevers in resonant bending. How this has been realized is explained in the following part. First, the used single setup parts and their capabilities will be introduced, and then the whole setup will be described. After that the testing processes and methodology of how the lifetime can be determined will be explained.

3.3.1. Hardware for Fatigue Setup

The physical cable connections of all hardware parts described below are displayed in Fig. 3.11. Exciting a Si cantilever at its resonant frequency is the working principle of the setup. For this purpose a piezoelectric actuator (P-842.60, PI Physik Instrumente GmbH, Germany) with a travel range of 90 µm is mounted. It is controlled via an amplifier (E-505.60, E-501.00, PI Physik Instrumente GmbH, Germany). With the forward and backward motion of the piezo, the sample cantilever can be excited at its resonant frequency.

To control the sample displacement amplitude and surface quality a red diode laser (NT57-113, Edmund Optics, Germany) is used. It has a wavelength of 635 nm and a power of 30 mW. The laser beam intensity is reduced by a non-reflective neutral density filter with optical density of 1.3 (NT63-381, Edmund Optics, Germany) which is sufficient enough for the setup. To guide the laser beam, a beam splitter (NT47-009, Edmund Optics, Germany) is placed in the optical path. Illumination is realized with a UV black light. Both, the laser and the black light, are turned on and off via a USB plug.

For the optical analysis an area detector (SPOTANA-9, Duma Optronics, Israel) is used, which is a position sensitive detector. It has a 9 x 9 mm chip and measures the x- and y-position as well as the intensity of the reflected laser beam. A precision longpass filter (NT62-978, Edmund Optics, Germany) was fixed in front of the chip to only get the signal of the laser beam. For further optical analysis a camera (PL-B782, PixeLink, Canada) is present, which has a 6.6 MPixel CMOS chip and an attached 1:1 objective.

Data acquisition of the piezo actuator and the area detector was performed by a USB data acquisition box (NI USB-6251, National Instruments, USA) with eight inputs (16-bit and 1.25 MS/s).

To be able to position and scan the sample, two micro stepper motors (M-230.25, Physik Instrumente GmbH, Germany) with a resolution of 0.05 mm are mounted at the stage. They are driven by a two axis PCI controller card (C-843, Physik Instrumente GmbH, Germany).

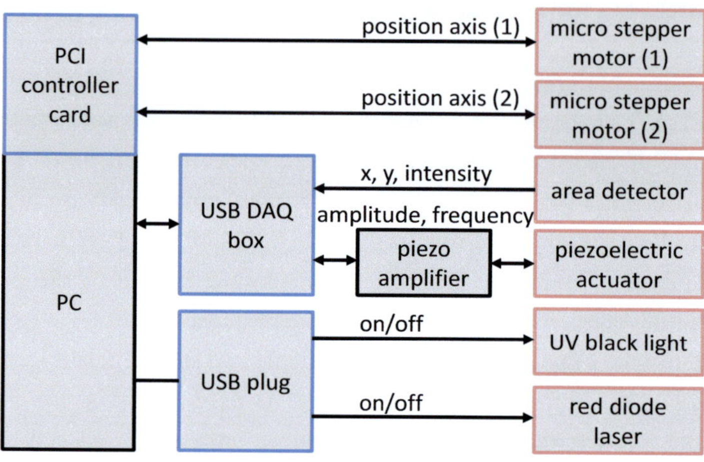

Fig. 3.11: Schematic of the physical cable connections of the hardware parts and the obtained data.

All the hardware parts are now assembled to the cantilever fatigue setup. A schematic of the setup is displayed in Fig. 3.12. For the sake of simplicity and to point out the working principle the schematic is not to scale. In the center of the image there is the cantilever array with different masses attached to the cantilever free ends. The array is glued on a stub which is screwed onto the piezoelectric actuator behind the sample. The forward and backward motion of the piezo excites a cantilever at its resonant frequency in an out-of-plane bending mode. This whole arrangement sits on a stage, which can be moved in x- and y-direction by micro motors (not displayed in the image). Perpendicular to the sample array is the laser. The laser beam is directed onto the sample surface. The reflection of the laser beam from the sample surface is deflected with the beam splitter by 90 ° onto the area detector. The area detector can measure the displacement of the cantilever as well as the intensity of the reflected laser beam. Directly above the sample, the camera is mounted. With this the maximum displacement amplitude u_z^{max} of the excited cantilever can be monitored. The whole setup is placed in a box to minimize influence from ambient light. In the box there is a UV-black light for a better detection of the displacement of the cantilever ends. A real image of the setup is shown in Fig. 3.13.

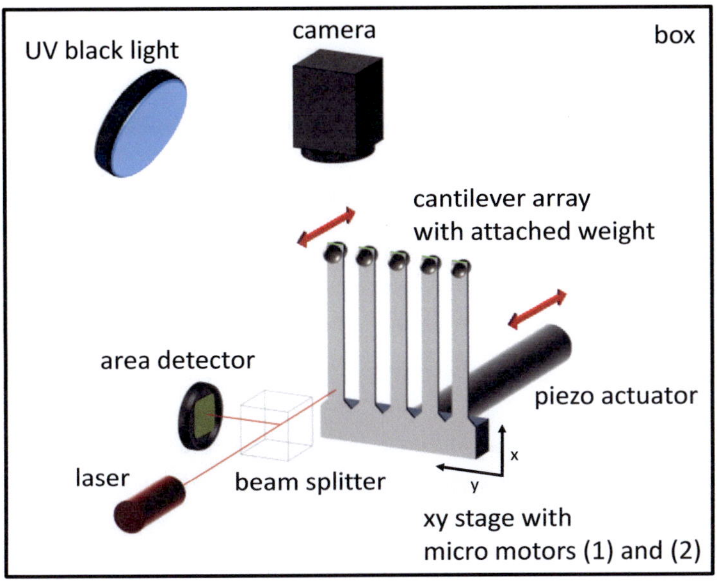

Fig. 3.12: Schematic of the cantilever bending fatigue setup and the working principle (not to scale).

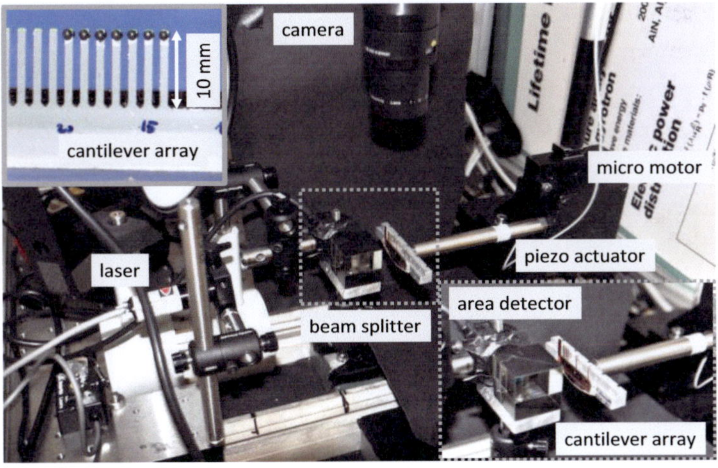

Fig. 3.13: Image of the thin film fatigue setup. The insets show the cantilever array in more detail.

3.3.2. Testing Procedure and Software Implementation

In this chapter, the testing procedure beginning from the sample preparation until the actual experiment will be explained. A short overview of the software implementation is given.

First, an array of cantilevers (see inset Fig. 3.13) is fixed on the sample holder stub with super glue. After a drying time of one day the free ends of the cantilevers are painted with a fluorescent green color in order to be able to measure the displacement with the camera under UV-black light. To manipulate the resonant frequency of the cantilevers, masses with different weights are glued to the cantilever ends. Two Pb spheres of the same weight are glued on the front and back side at the free end of a cantilever. Important to note is that the total weight of the two spheres has to be in the range between 9 – 22 mg (see chapter 3.2.4). To avoid having two cantilevers with the same resonant frequency the attached masses must have at least a difference of 1 mg. After drying again for a day the whole sample is screwed onto the piezo. Once the alignment is done the experiment can be conducted.

Following the operation chart in Fig. 3.14 the different testing programs (indicated by a program number) are implemented in LabView (Version 8.5, National Instruments, USA). For further information the program structure of each LabView program is displayed in Fig. 3.15. Generally, the testing routine consists of three main steps, namely scanning, sweeping and cycling.

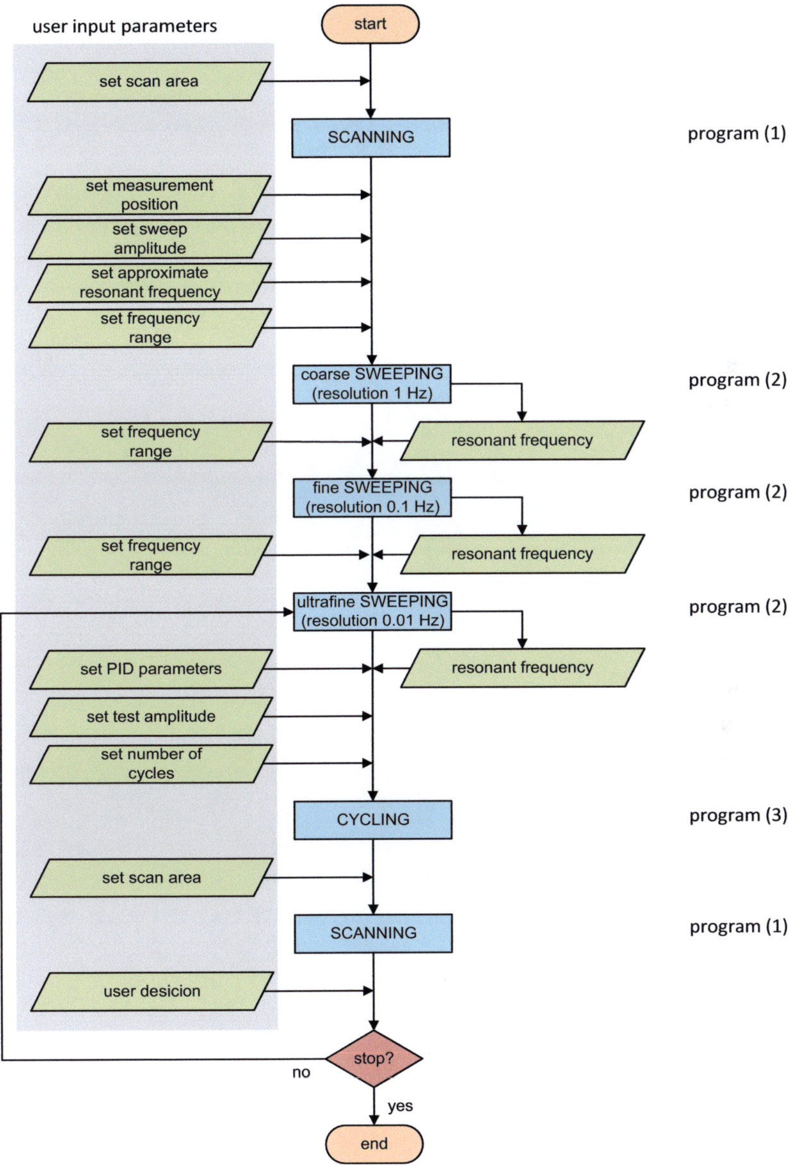

Fig. 3.14: Dynamic testing procedure scheme with user input parameters and the connection of the implemented testing programs.

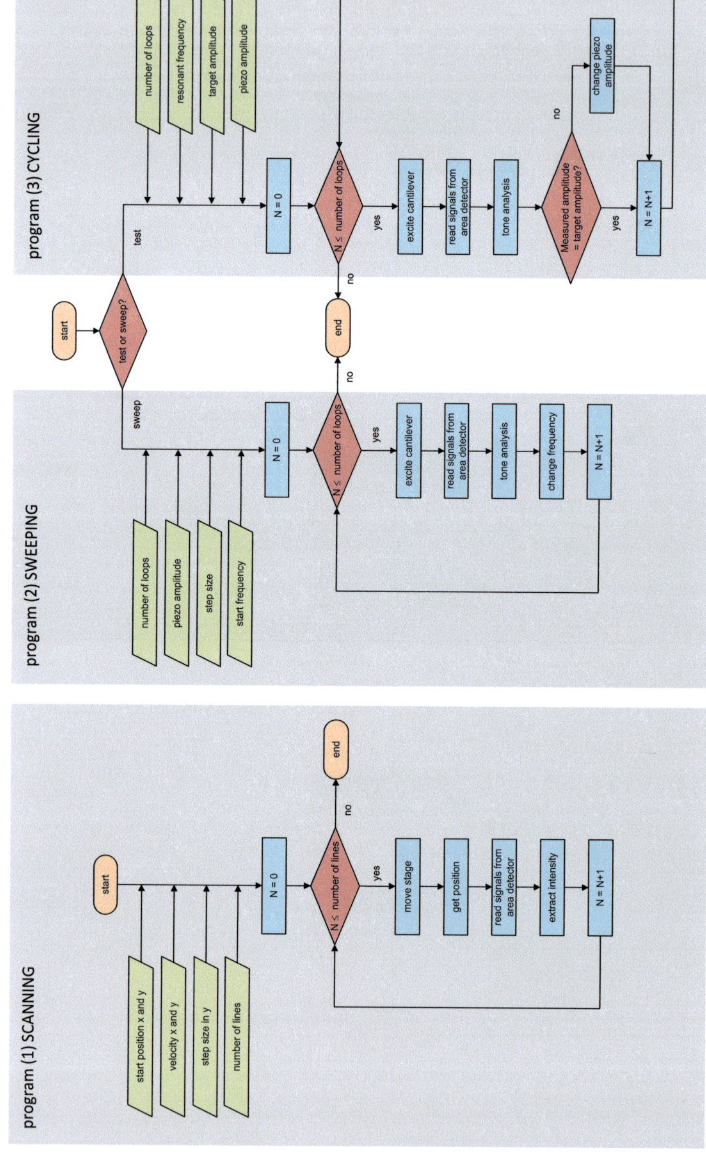

Fig. 3.15: Testing procedure scheme of the three testing programs for scanning, sweeping and cycling.

At the very beginning the sample surface is scanned in its initial state. During scanning the sample is continuously moved in the x-direction and stepwise in the y-direction, like a longitudinal line profile. Simultaneously, the reflected intensity from the sample surface is recorded by the area detector and can later be correlated to the actual position on the sample. The velocity for the x-direction is 0.1 mm/s, which is slow enough to measure the intensity with an adequate resolution. The y-position, with a step size of 0.2 mm, is kept constant while scanning in the x-direction.

The cantilever is then positioned in a way that the laser beam reflects from the center of the cantilever roughly 1 mm away from the 45 ° shoulder. The next big step is to find the resonant frequency of the desired cantilever, which is done by sweeping. The resonant frequency can be roughly approximated by the measured thickness of the cantilever t_{sub} and the attached weight from the FE simulation performed and plotted in Fig. 3.8. The first sweep step covers a range of $\Delta f = 200$ Hz around the approximated frequency. The frequency of the piezo is stepwise reduced from the maximum to the minimum frequency with a constant excitation amplitude of 0.01 V = 0.09 µm. The excitation amplitude is chosen to be small on purpose, so that the sweeping does not induce any damage beforehand. During the sweeping through the frequency range the amplitude of the reflected laser beam is measured by the area detector. A maximum in the measured amplitude indicates the resonant case. Now iteratively the range around the maximum is reduced to $\Delta f = 20$ Hz and finally $\Delta f = 2$ Hz. In the end the resonant frequency can be determined with a precision of 0.01 Hz. Typically the full width half maximum (FWHM) of such a resonant peak is 0.5 Hz. Examples of a fine and ultrafine sweep are displayed in Fig. 3.16.

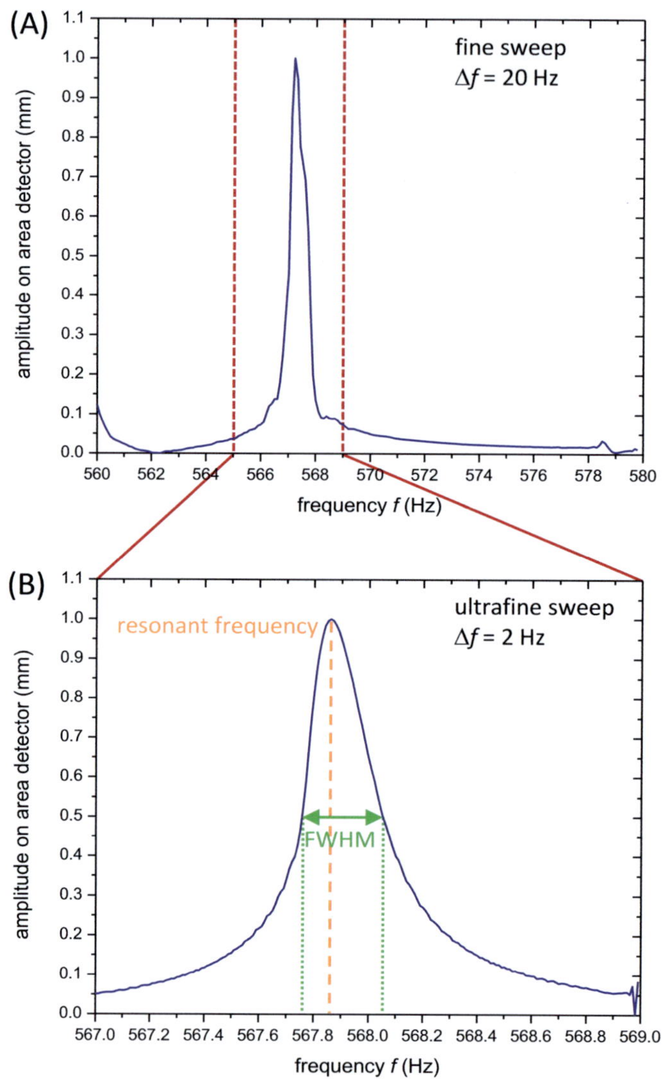

Fig. 3.16: Amplitude on area detector versus the frequency for (A) a fine sweep and (B) an ultrafine sweep. The resonant frequency is indicated by the maximum aplitude.

Before testing, an appropriate maximum strain amplitude ε_a^{max} has to be chosen. At the resonant frequency f_{res} the excitation amplitude of the piezo is steadily increased while the amplitude on the area detector and the displacement of the cantilever end is monitored with the camera. By measuring the maximum displacement u_z^{max} with the camera and calculating the maximum strain amplitude with Eq. 3.8 at the same time, the piezo amplitude is increased until the desired maximum strain amplitude is reached. The final testing excitation amplitude refers to the measured amplitude on the area detector, which becomes the PID control parameter. By regulating the piezo excitation amplitude with the measured amplitude on the area detector it is ensured that the maximum strain amplitude is kept constant. For the safety of the sample, a maximum piezo excitation amplitude cannot be exceeded, and therefore protects the cantilever from breaking.

With all parameters found, the cycling step can be conducted. The piezo excites the sample at its resonant frequency at the desired excitation amplitude for a certain number of cycles. To double check the displacement amplitude, the camera takes images of the vibrating cantilever during testing. After the desired number of cycles is reached, the test stops automatically.

Now, this procedure starts over again: The cycled cantilever is scanned for its reflected intensity on the sample surface. Sweeping for the resonant frequency records a possible shift and then another cycling test is started. The test is interrupted after certain steps of cycling (e.g. 10^5, 10^6, 5×10^6, 10^7, 2×10^7, 5×10^7, 10^8), scanned for damage and swept for a change in resonant frequency.

3.3.3. New Approach to Determine the Lifetime

Due to the fact that the thin film on top of the Si cantilever is subjected to fatigue and not the substrate itself, the situation differs compared to conventional bulk fatigue testing. A single crack in the thin film will not lead to fatal failure, as the thin film is still attached to the intact substrate. This implies that another failure criterion than cracking has to be established for assessing the failure of the thin film.

As explained before, it is expected that the film is first damaged at the fixed end and that then the damage front is likely to move from the shoulder along the cantilever to the free end. By scanning the thin film surface with the laser and measuring the reflected intensity, the damage can be detected optically. The cycling induces extrusion formation, which roughens the surface and therefore disperses the laser beam and reduces the measured reflectivity. An example of such a reflectivity scan is displayed in Fig. 3.17 in the case for a 1 μm Cu thin film after 10^8 cycles with a maximum strain amplitude of $\varepsilon_a^{max} = 0.16$ %. The reflectivity scans are contour plots, where the x- and y-direction are the axes and the reflectivity of the laser beam is displayed in a color code. For a better comparison to other scans the scale of the reflected laser beam is normalized with respect to the intensity in the center of the cantilever at $x = 8$ mm. The color code for the reflectivity has to be read the following way: red means 100 % reflectivity, which indicates a fully reflected laser beam (undamaged region), whereas dark blue means 0 % reflectivity and therefore indicates a strong dispersion of the laser beam in the damaged region. Every color in between indicates a gradual decrease in reflectivity.

Fig. 3.17 (A) shows the initial state of a thin Cu film. The shape of the cantilever can clearly be depicted by the red area. The shoulder of the cantilever is at position $x = 0.8$ mm and the attached mass is located at position $x = 9.8$ mm. In Fig. 3.17 (B) the intensity scan of the same cantilever is displayed after 10^8 cycles. For a better comparison the original shape is marked by a dotted line. Half of the cantilever is still red, which means that this region is undamaged. From position $x = 4.5$ - 1.1 mm, a gradual decrease of reflectivity from yellow to green to light blue can be identified, which indicates damage of the thin film surface. It should be noted that the cantilever itself is still fully preserved. Supporting the argument that reflectivity changes in damaged regions, in (C) a micrograph of the cantilever is shown. Near the shoulder of the cantilever small features at the thin film surface can be identified. These features correspond to the region where the reflectivity decreases and indicate that fatigue damage was induced during cycling.

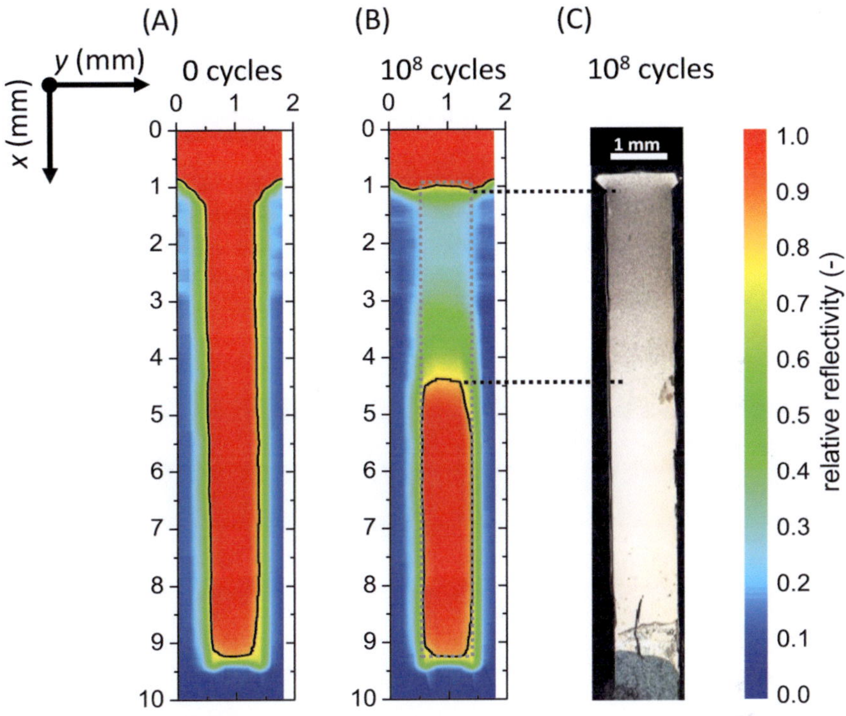

Fig. 3.17: Comparison of reflectivity scans in the initial state (A), after 10^8 cycles (B) and an optical micrograph (C). The decrease in reflectivity in (B) corresponds to the damaged region in (C).

Based on this general trend, failure has to be defined by the optical decrease in reflectivity. In Fig. 3.18 the relative reflectivity of an Al thin film after 10^8 cycles is plotted along the x-position of the cantilever in the center of the sample, indicated by the blue dotted path. Between $x = 1.5 - 2.5$ mm the reflectivity displays a minimum and increases until x = 7.5 mm. A certain amount of damage may be tolerable. Therefore, the basic definition of failure is a loss of 20 % in reflectivity, but other threshold values depending on the application can be easily applied. Regions of the cantilever that are below a reflectivity of 0.8 are seen as damaged.

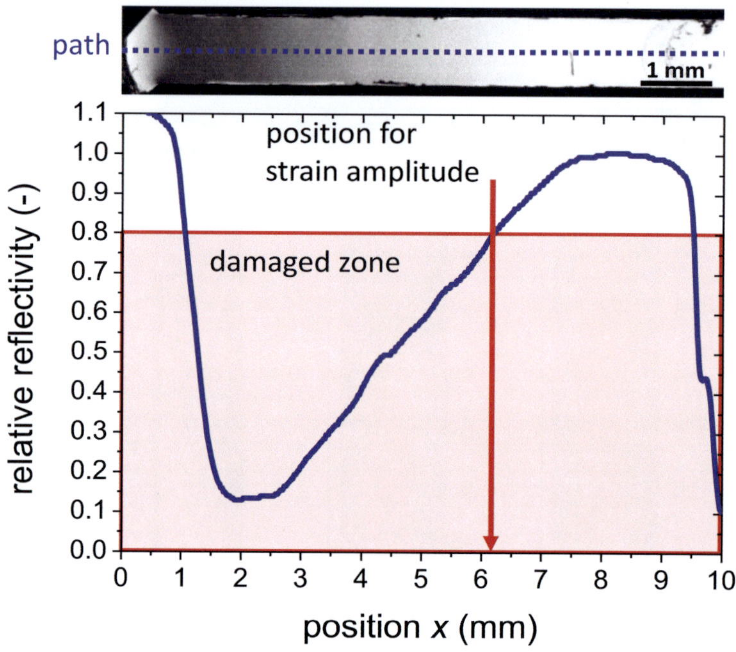

Fig. 3.18: Relative reflectivity versus the position x along the cantilever on the center path. Regions with less than a relative reflectivity of 0.8 are considered to have damage on the thin film surface. The position, where the failure criterion is fulfilled can be correlated to the strain amplitude acting in the thin film during cyclic bending.

The strain amplitude ε_a acting in the thin film is linearly correlated to the position on the cantilever along x (see chapter 3.2.3). With the defined failure criterion the position of the damage front can be identified and the corresponding local strain amplitude can be calculated. The testing procedure is done in a stepwise interruption and scanning of the sample surface (see chapter 3.3.2). This has the advantage that the moving damage front is monitored with respect to the cycle number.

With this technique one lifetime diagram can be extracted from testing only one cantilever in the following way (Fig. 3.19): All the reflectivity scans of an Al thin film surface with the referring cycle numbers are displayed in (A). The particular position of the fulfilled failure criterion of 0.8 is indicated by the black contour line. Following the arrows, the position is correlated to the corresponding local strain amplitude by using data from FE simulation, shown in (B). Finally, the strain amplitude is plotted versus the matching cycle number. All data is processed that way to obtain the lifetime data of the samples.

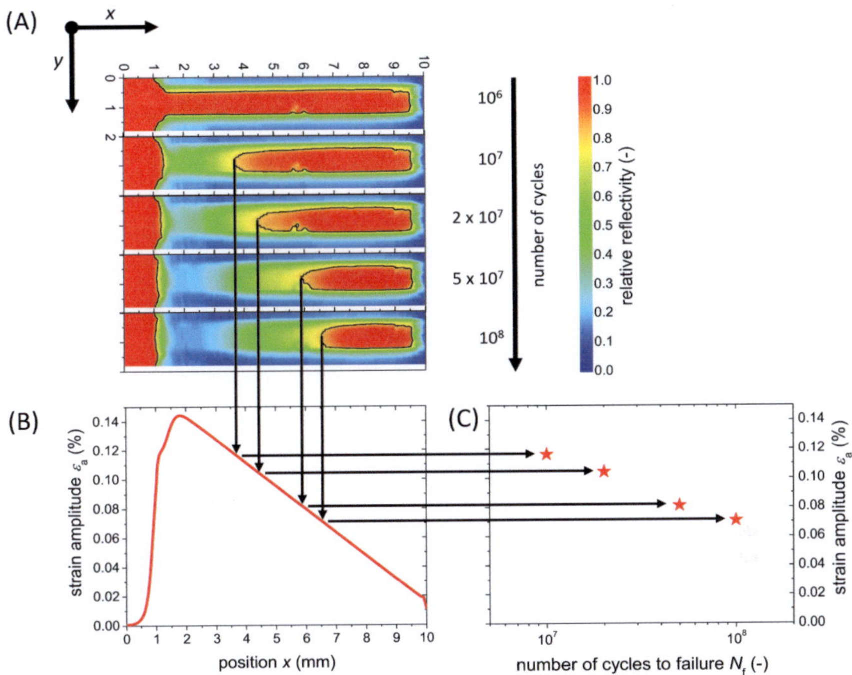

Fig. 3.19: (A) Reflectivity scans at different cycle numbers with indicated failure criterion contour at a relative reflectivity of 0.8. To obtain a lifetime diagram the position of the damage front is correlated to the local strain amplitude at this position (B) and then plotted versus the number of cycles (C).

Another way to clarify the procedure is displayed in Fig. 3.20 for a Cu thin film. Here, the strain amplitude ε_a on the left axes and the corresponding position x on the right axes are plotted versus the number of cycles to failure N_f. In the background three representative reflectivity scans after 5 x 10^6, 2 x 10^7, and 10^8 cycles are displayed exemplarily with the corresponding failure contour line. The data points display the evolution of the damage front along the cantilever and the resulting lifetime diagram. With only one sample, the data for a whole lifetime diagram can be gathered. Therefore, this setup works in a high-throughput manner.

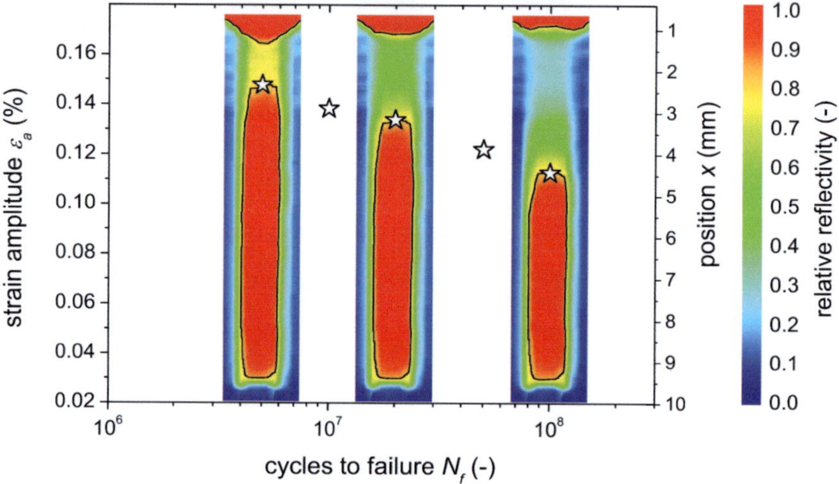

Fig. 3.20: Lifetime diagram with the strain amplitude ε_a on the left axes and the corresponding position x on the right axes versus the number of cycles to failure N_f. In the background the corresponding reflectivity scans are plotted to exhibit the evolution of the damage front. The black line corresponds to a failure criterion of 0.8 in relative reflectivity.

3.4. Discussion

In this section the presented methodology for the novel resonant bending fatigue setup for thin films will be discussed with respect to general aspects. In the following paragraph an error analysis of the calculated local strain amplitude is presented. After that the results from the FE simulations will be discussed. Then, other implementations of the setup are the focus of discussion as well as further improvements and suggestions for additional developments. The chapter will be closed by a critical look on the lifetime criterion used in this thesis.

3.4.1. General Aspects of the Cantilever Bending Setup

The presented implementation for a resonant cantilever bending setup for thin film fatigue testing is a powerful tool to obtain lifetime diagrams in a short time with a low number of samples. The setup can be used in a high-throughput manner, systematic fatigue analysis of pure element thin films with different thicknesses and microstructures are possible as well as the investigation of whole alloying systems or multilayers. Therefore, the experimental principle can be used as a combinatorial high-throughput experiment. Any thin film that has a reflective surface and is more pronounced to fatigue than the substrate can be tested. Various combinations of substrate and thin film materials allow for studying the influence of the interface properties on the fatigue behavior.

In this work Si was chosen as substrate material because of the relevance to microelectronic devices and its material properties. In principle other substrate materials e.g. ceramics or high strength steels are possible. Besides costs and complexity of the fabrication process for the sample shape, the material has to meet certain properties. It is favorable if the substrate has a high tensile strength, as this gives the limit for the fatigue lifetime of the substrate. Furthermore, a low Young's Modulus would be good in terms of reducing stresses in the substrate, but this is typically contrary to also have a low damping coefficient.

A big advantage of the technique is that it is a non-contact method and can be automated. Moreover, it is very sensitive to surface changes, which is favorable for high resolution measurements. On the other hand, it has to be kept in mind that for the same reason processes below the surface may not be captured. Therefore, additional microstructural investigations are recommended.

As mentioned before, the principle itself can be used at different dimensions. The ratio of film thickness and substrate thickness is larger than 1:150 to ensure the thin film approximation and a uniform strain gradient in the thin film along the cantilever. The usage of other ratios for thicker films or even for a bulk material is also possible but additional strain gradients over the cross section of the cantilever have to be taken into account. This also applies for up or down scaling of the cantilever size, if the ratio of film to substrate thickness changes. Down scaling might be feasible and interesting for an *in situ* implementation and observation of extrusion formation, e.g. in an SEM. It has to be kept in mind that changing the dimensions of the sample also changes the resonant frequency changes and other excitation devices have to be used. For this, it is recommended to conduct an FE analysis for such cases, in order to calculate the maximal strain amplitudes and gradients.

3.4.2. Error Analysis

The calculation of the local strain amplitude at the position of the damage front is based on FE simulations, fitting an analytical solution to the strain distribution, and the experimental measurements. All these steps introduce individual errors which have to be considered in the error analysis. The combined error is approximated by the method of error propagation. The equation for the local strain amplitude is partially differentiated and multiplied with the uncertainty for each variable.

First, the error of the fitted function for the calculation of the maximum strain amplitude ε_a^{max} is calculated. The basic equation is displayed in Eq. 3.8 and the uncertainty $\Delta\varepsilon_a^{max}$ can be calculated in the following way:

$$\Delta\varepsilon_a^{max} = \frac{\partial\varepsilon_a^{max}}{\partial u_z^{max}} \cdot \Delta u_z^{max} + \frac{\partial\varepsilon_a^{max}}{\partial t_{sub}} \cdot \Delta t_{sub} \qquad \textit{Eq. 3.9}$$

resulting in:

$$\Delta\varepsilon_a^{max} = \left(A + Bt_{sub}\right) \cdot \Delta u_z^{max} + \left(Bu_z^{max}\right) \cdot \Delta t_{sub} \qquad \textit{Eq. 3.10}$$

where Δu_z^{max} is the standard deviation of the measured displacement amplitude of several images and $\Delta t_{sub} = 10$ µm is the uncertainty of the measurement of the Si substrate cantilever thickness. The average error for ε_a^{max} is calculated to be 4 %.

The local strain amplitude ε_a^{loc} at position x can be calculated by a linear fit of the FE data with the abscissa C and slope D

$$\varepsilon_a^{loc} = C + D \cdot x \qquad \textit{Eq. 3.11}$$

By differentiation, the error propagation can be calculated:

$$\Delta\varepsilon_a^{loc} = \frac{\partial\varepsilon_a^{loc}}{\partial C}\cdot\Delta C + \frac{\partial\varepsilon_a^{loc}}{\partial D}\cdot\Delta D + \frac{\partial\varepsilon_a^{loc}}{\partial x}\cdot\Delta x \qquad \textit{Eq. 3.12}$$

resulting in

$$\Delta\varepsilon_a^{loc} = (1)\cdot\Delta C + (x)\cdot\Delta D + (D)\cdot\Delta x \qquad \textit{Eq. 3.13}$$

with ΔC and ΔD correspond to the error from ε_a^{max} and $\Delta x = 100$ µm. Therefore, the overall error for the local strain amplitude ε_a^{loc} at the position of the damage front is around 7 %. The error of the strain amplitude is later calculated individually for each data point and is displayed as error bars.

Another error is introduced by the fact that the setup needs a certain number of cycles to reach the given piezo excitation amplitude. Furthermore, the low damping coefficient of the Si cantilever increases the time the sample is exposed to a high excitation amplitude after the piezo actuator is switched off. The error in number of cycles can be estimated as follows: after the desired cycle number is reached, the test is stopped automatically. Experience shows that the cantilever comes to a full stop after up to 10 s at a resonant frequency of 700 Hz, which would result in 7,000 extra cycles. For the lowest first cycle step of $N = 500{,}000$ that would be an error of 1.4 %. Since the displacement amplitude drops rather fast at the beginning, the number of cycles which might have an impact on fatigue lifetime can be estimated to be less than 1 %. With increasing cycle numbers for the following cyclic loading steps the error is even less. The same is valid for the beginning of the measurement. Therefore, the error made by counting the cycles is neglected.

3.4.3. Finite Elements Simulation Results

The FE simulations presented here helped to analyze the specific experiments conducted in this thesis. For other cases the FE parameters like element type and size, the used analysis type and the influences of geometry or boundary conditions have to be adapted. Here, a 'modal' analysis was chosen which accepts linear elastic properties of the involved materials but allows to consider complex geometries compared to the analytical solution.

Nevertheless, the presented FE simulations can be further improved by adding an actual thin film to the substrate and calculating the strain and stresses in the thin film directly. However, this requires more calculation power and detailed information on thin film material data, which may not be available. Additionally, interface properties have may become important when the strain is not fully transferred from the substrate to the film.

The mass was simplified as a cube, to vary the weight of this mass with the choice of according densities. The shape of the weight has little impact on the local strain amplitude, as long as it is placed at the same position at the end of the cantilever. Further accuracy in calculation is achieved if instead of an 8 node 3 D element, a 20 node 3 D element is used to conduct a geometrically non linear analysis.

The results from the ‘modal’ analysis can be used as input for a ‘transient’ analysis if the behavior with time should be calculated. This would allow to take material damping into account, which could improve the strain amplitude calculation. But as it was pointed out before, as the damping of the system is rather low, such calculations were discarded.

3.4.4. Experimental Outlook for Next Generation Fatigue Setup

The current PID control of the setup monitors the excitation amplitude. This could be enhanced by additionally monitoring the resonance frequency by controlling the phase angle between the actuator and the vibrating cantilever [99]. The resonance peak is rather sharp and therefore the resonant system is prone to react sensitively to the fatigue induced damage in the thin film. Nevertheless, the thin film contributes only little to the stiffness of the cantilever and therefore the amplitude control seems to be a good solution for the presented system. If the thickness ratio between film and substrate increases, an improved frequency control might be mandatory.

Instead of using a reflected laser beam and an area detector also a high speed camera or a microscope could monitor the surface quality of the thin film. In order to ensure a sufficient resolution, the camera field of view has to be reduced, and possible information might be lost. On the other hand, this would open up the possibility to use digital image correlation (DIC) [26] to measure displacements or strains directly on the sample surface.

As already described in [83, 100] the sample could also be observed from the side to get the full information on the bending contour of the cantilever. The disadvantage for a high-throughput approach is that the camera has to be rearranged for every cantilever, additionally the other cantilevers of the array are in the line of sight.

Transferring the setup into an SEM [101] or a synchrotron and modifying it accordingly would make it possible to do *in situ* thin film fatigue testing where the damage formation could be monitored while it happens. Installing the setup into a climate chamber would allow studying the influence of temperature, humidity and atmosphere.

3.4.5. Comparison of Failure Criteria

For the presented fatigue setup the relative reflectivity of the thin film surface along the whole cantilever is used to identify the damage front and failure criterion. During cycling, the surface of the thin film changes due to surface roughening by extrusion formation. With increasing roughness, the laser beam is dispersed and thereby the reflectivity, which is measured with the area detector, decreases. Another optical criterion was used by Eve et al. [77]. They used the stray light of biaxial tested Au thin films on a polymer substrate. For the optical detection, photo sensors are placed around the sample and a decrease in output voltage corresponds to failure. This kind of failure detection represents only a global criterion for the whole sample and is not position sensitive as the presented method in this thesis.

Other failure criteria were proposed, depending on the setup and the boundary conditions an appropriate failure criterion has to be chosen (see Table 2.1).

Read et al. [74] implemented a tension-tension fatigue setup for freestanding Cu thin films, which is controlled in closed loop by measuring the displacement. Failure is defined, if the maximum capacity of the driving piezo stack is reached to maintain the load amplitude after elongation of the sample. Therefore, the lifetime is reached when one crack becomes long enough for fatal failure. Schwaiger et al. [90] tested Cu thin films on a polyimide substrate in a cyclic tension compression test. They defined a further decrease of the mechanical energy loss after a steady-state period as failure. This criterion is based on an integral signal, reflecting the defect accumulation in the thin film. Another possibility for these kinds of tests has been applied by Wang et al. [80]. They interrupted the fatigue tests periodically and investigated the damage evolution using scanning electron microscopy. The defined failure criterion in these observations is the number of cycles, where the evolution of extrusions or cracks saturates. This criterion might be comparable to the reflectivity criterion used in this study. If testing Ag thin films on SiO_2 micro cantilevers loaded by a nanoindenter [75], a stiffness drop after a beforehand reached plateau is regarded as an indication of damage formation. This criterion seems to be more similar to the criterion of Read et al. [74] as the impact of one large crack can be enough to reach the lifetime limit. Experiments on the resistivity change of Al, Cu [81] or Ag [86] thin films on polymer substrate during cyclic loading in the LCF regime introduced another failure criterion. As damage occurs, the resistivity increases. Although, Sun et al. [81] and Sim et al. [86] used a similar testing setup, they defined different failure criteria. While Sun et al. [81] define failure at the intersection of two linear regimes of the resistivity curve and attributed this to crack nucleation and growth. Sim et al. [86] simply use an absolute resistivity increase of 25 %. The latter argue that their criterion gives more reasonable results as the identification of the linear regimes is difficult. On the other hand, their criterion lacks of a physical explanation. In the case of thermally cycled Cu lines the failure is defined as an electrical open [76]. This seems to be a fatal criterion as well, because one crack is sufficient to lead to failure.

It can be noted that the choice of failure criterion is case-specific and strongly depends on the testing setup and samples as well as the intended application of the thin film. Nevertheless, most criteria from literature have in common that no local damage information can be extracted. The innovation and advantage of the here introduced reflectivity failure criterion is that the evolution of damage can be monitored along the cantilever at different cycle numbers.

Depending on the stiffness of the substrate the stress relaxation in the region around the crack is different. For a polymer substrate the stress release is large and the crack can open easily and even propagate. In the case of a stiff Si substrate the region is small and no crack propagation takes place. Therefore, specifically in this study the processes of crack nucleation and growth as observed in bulk material are not of particular interest, but the overall evolution of damage formation and features.

The failure criterion for a relative reflectivity of 0.8 presented in chapter 3.3.3 is for the moment a more or less arbitrary choice. However, the criterion can easily be varied in a post process from a relative reflectivity 0.95 down to 0.5 as all the data is measured at the same time and available after the test. This allows the definition of an upper and lower boundary of the fatigue limit and can be varied accordingly with respect to the future application of the thin film. If the application is a high reflective optical sensor, then a relative reflectivity of 0.95 is suitable as failure criterion. But if an application depends on conductivity, even a relative reflectivity of 0.5 can be tolerated as electron percolation is still available throughout the cross section. As the reflectivity is surface sensitive, the damage throughout the thickness might not be severe. In the case that the directly measurable lifetime criterion is not adequate for an application, the measurable surface damage has to be correlated to, e.g., the resistivity of the thin film. For this further measurement methods have to be added to the setup or carried out after the experiment, e.g. a small scale 4-point resistivity measurement with high local resolution.

4. Experimental Methods for Sample Fabrication and Characterization

This chapter on experimental methods for sample fabrication and characterization will give a brief overview how samples were fabricated. First, the cantilevers were etched and then a thin film is deposited by magnetron sputtering. The microstructure of the thin films was characterized by optical microscopy techniques. Furthermore, other characterization techniques will be presented.

4.1. Sample Fabrication

The thin film material system was chosen due to the high interest in applications. The substrate is Si as it is stiff and responds in a linear elastic fashion at the strain amplitudes which are expected in fatigue experiments. In the here envisioned experiments, Si is not expected to fatigue (see e.g. [102, 103]. Furthermore, Si is widely used in the microelectronics area and is the standard substrate for thin films. Therefore, structuring Si by an etching technique is a fairly known process and in principle all kinds of geometries can be realized. Cu and Al were chosen as thin film materials since both are of high interest in the microelectronics industry as they provide high conductivity and a high corrosion resistance. Cu thin films were studied in the recent past for its small scale mechanical properties [42, 104-106] as well as Al [38, 107, 108].

4.1.1. Fabrication of Si Cantilever Arrays

The Si cantilever samples were etched and the thin films deposited at the Department "MEMS Materials/Institute of Materials" in the group of Prof. Alfred Ludwig at the Ruhr University in Bochum. To achieve a high-throughput experiment, several cantilevers were fabricated in arrays of either 16 or 32 cantilevers (Fig. 4.1). Using the space on a 4 inch Si wafer efficiently, the 32 cantilevers (arrays 2 and 3) were arranged in the center, and the 16 cantilevers (arrays 1 and 4) near the outside. The desired geometry of each Si cantilever is a length of 10 mm, a width of 1 mm and a thickness of 200 µm. To avoid the influence from neighboring cantilevers, the gap between the cantilevers was set to a distance of 1 mm. The wafer was prepared with a masking process and subsequently the arrays were fabricated with a wet etching process.

Double-side polished 4 inch (100) Si wafers with a thickness of 500 µm were thermally oxidized, resulting in a 300 nm thick silicon oxide layer. A 100 nm silicon nitride layer was then deposited by low-pressure chemical vapor deposition, which was used as the mask for etching. Additionally, the silicon nitride film compensates stresses induced by the oxide layer, and it also serves as a diffusion barrier to prevent reactions between Si and the later deposited thin films. Photolithography was performed on the backside with the array mask using a positive photoresist (AZ1518, AZ Electronics, Germany) and a mask aligner (SUSS MA6, SÜSS MicroTec AG, Germany). The cantilever arrays were aligned in the [100] direction to produce cantilevers with straight sidewalls. The local removal of silicon nitride and silicon oxide was performed by reactive ion etching, and the photoresist was removed in oxygen plasma. A KOH wet etch process (30 % KOH at 80 °C) was applied for the final release of the cantilevers. Trenches were etched on the wafers to facilitate the separation into four cantilever arrays (see also [109]).

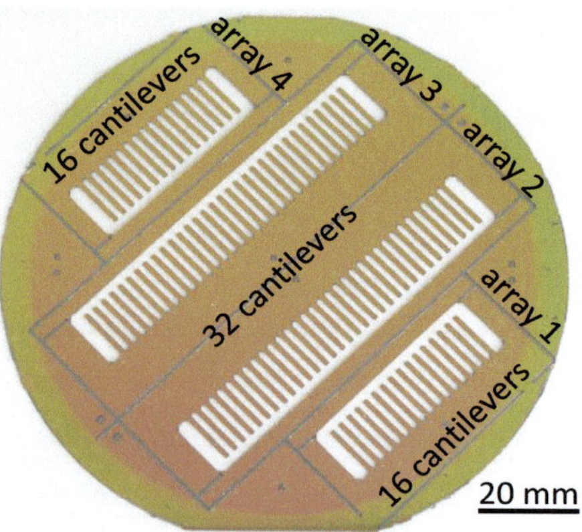

Fig. 4.1: Image of the front side of a 4 inch Si wafer with the etched cantilever arrays.

Due to improper etching processes, the cantilevers displayed etch pits and grooves at the backside or side walls (Fig. 4.2). As Si is a brittle material a small defect near the surface can lead to fatal failure of the cantilever, before the desired strain amplitude could be reached. The highest strain amplitude tested without breaking the Si cantilever was around 0.23 %, which equals a stress of around 350 MPa. This is far below 3 GPa, that was tested to be a limit of fatigue strength for 10^{11} cycles [102]. Table 5.1 therefore only lists the successfully tested cantilevers. The failure rate of the Si cantilevers per array was sometimes quite high (up to 2/3) depending on the Si wafer quality and the etching process during fabrication. For future sample preparation, it is desirable to improve the etching process to avoid any defects.

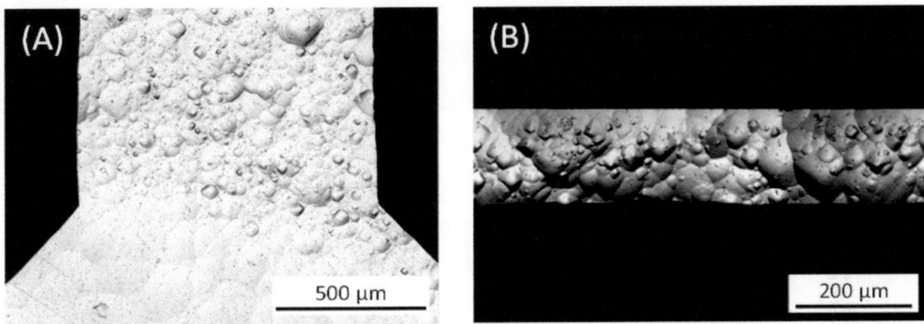

Fig. 4.2: Micrograph of etch pits on the back side (A) and side wall (B) of a Si cantilever.

4.1.2. Thin Film Deposition

Thin films with a thickness of approximately 1 μm were deposited on the etched cantilever arrays by using a combinatorial magnetron sputtering system (DCA Instruments, Finland). Generally, the deposition was carried out in an Ar plasma, the Si substrates were not heated, kept at ground potential and note rotated. All other parameters were adjusted according to Table 4.1. The different targets of 100 mm diameter were positioned 130 mm above the substrate. The base pressure prior to deposition varied between 1.5 x 10^{-6} Pa and 4.7 x 10^{-6} Pa, while the pressure during deposition was adjusted between 0.27 Pa and 2.00 Pa. The sputter power was kept between 150 W and 200 W either DC (direct current) or RF (radio frequency). To improve the adhesion of Cu films to the substrate, an RF bias with a power ranging from 25 to 50 W was applied (see also [109]). To obtain a different microstructure some Al and Cu thin film arrays were annealed afterwards for 1 h at 450 °C in a vacuum chamber. To study the influence of a seed layer, Cu thin films were deposited on a 10 nm Ti thin film.

Table 4.1: Sputter parameters for thin film deposition.

material	target purity (%)	base pressure prior to deposition (Pa)	Ar pressure during deposition (Pa)	sputter power (W)
Al	99.999	4.7×10^{-6}	0.67	200 DC
Cu	99.999	2.7×10^{-6}	0.27-2.0	200 RF
Ti	99.995	1.5×10^{-6}	0.67	150 RF
Cu on Ti	99.999	1.5×10^{-6}	0.67	200 DC

4.2. Methods for Microstructural Analysis

Three different types of microscopes were used and will be introduced in this chapter. Additionally, the data analysis techniques will be described.

4.2.1. Scanning Electron and Focused Ion Beam Microscopy

To gain a better understanding of the overall development of the microstructure and damage it is of interest to observe the microstructure on the surface as well as throughout the thickness of the thin films. For the necessary preparation, a dual beam microscope was used (Nova 200 NanoLab, FEI, USA). It combines a scanning electron microscope (SEM) and a focused ion beam (FIB) in one device. This has the advantage that surfaces can be monitored with the SEM and if necessary can be structured with the Ga^+ beam of the FIB. Generally, micrographs were taken with the electron beam at an accelerating voltage of 15 kV and a current of 0.58 nA either with an Everhart-Thornley Detector (ETD, for low magnification) or a through-the-lens Detector (TLD, for high magnification).

In this study the evolution of the damage structure along the cantilever is especially important. To carry out area analysis of the fraction of damaged area, micrographs were taken in three rows in the center of the cantilever 200 µm apart from each other. In the x-direction micrographs with 2000 x, 5000 x, and 10000 x magnification were taken every 500 µm (see Fig. 4.3 (A)). These micrographs (B) were transferred into a binary micrograph format (C) by choosing a grey value as a criterion for a damaged area (ImageJ, NIH, USA). Black indicates undamaged and white damaged areas. By measuring the fraction of the black or white area, the amount of undamaged or damage area can be obtained, respectively.

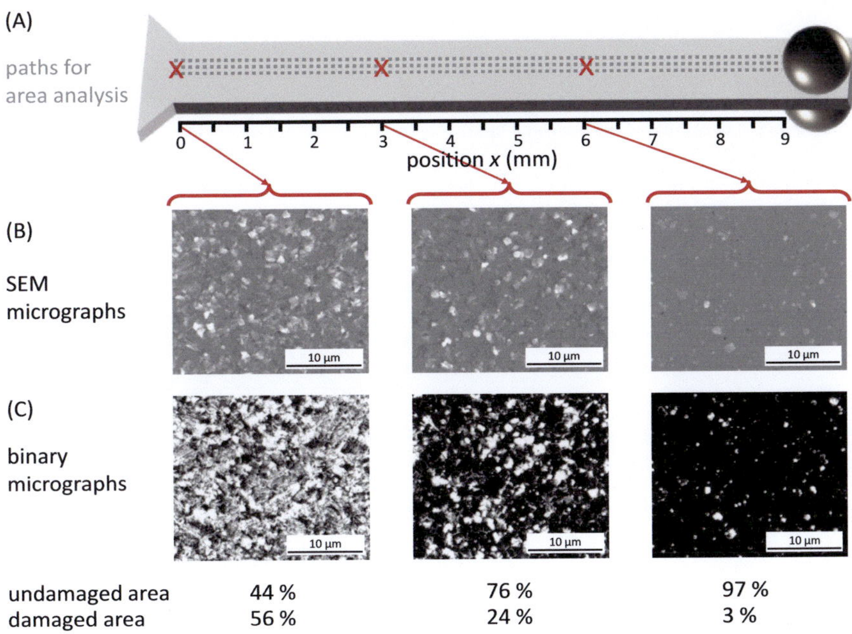

Fig. 4.3: Schematic for performing area analysis. (A) Cantilever with paths for imaging the thin film surface. (B) SEM micrographs taken at three exemplary positions of an Al thin film after 10^8 cycles. (C) Binarized micrographs with the fraction of undamaged and damaged area.

Additionally, the higher magnification SEM micrographs of the thin film surface were evaluated by a line analyses MATLAB code [110] to measure the size of damaged structures. In this analysis 30 lines were applied onto a micrograph. The damage structures were defined as individual damage features and not distinguished between different types. Grain sizes were measured from FIB cross sections. Fig. 4.4 (A) shows exemplarily a line analysis with individual damage features. In (B) a resulting cumulative distribution function plot is displayed. The median damage area size can be measured as indicated. The amount of the damage area features can then be calculated from the fraction of damaged area obtained from area analysis divided by its median size measured by line analysis.

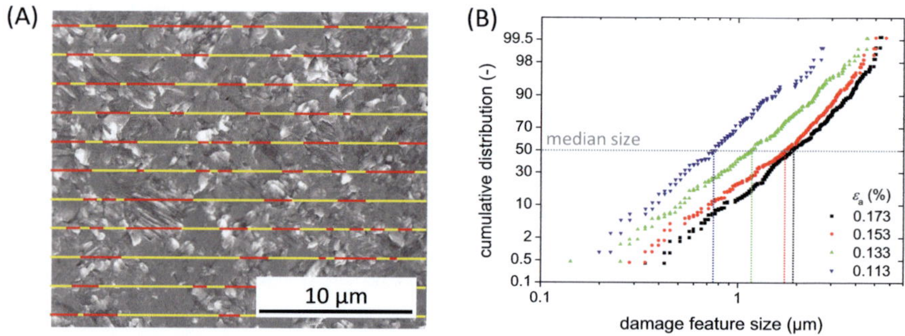

Fig. 4.4: (A) Schematic of line analysis. Damage features are extrusions or hillocks. The red fractures of the yellow lines indicate the damage features. (B) Example for a resulting cumulative distribution function plot, where the median damage area size can be measured as indicated.

To examine the microstructure and measure the grain size throughout the thickness of the thin film, cross sections were milled into the thin film by the FIB. First, a protective Pt layer was deposited with the FIB and a gas injection system inserted into the microscope chamber. An area of length x width x thickness = 10 x 1 x 0.5 µm was deposited with a current of 30 pA. Second, a staircase cross section was cut with dimensions of roughly 10 x 5 x 3 µm with a current of 1 nA. In a third step, the cross section quality was improved by several cleaning steps which were performed with reduced ion beam currents. For further reading the following book is recommended [111].

4.2.2. Transmission Electron Microscopy

To be able to analyze the fatigue induced dislocation structures, transmission electron microscopy (TEM) was performed. TEM samples were prepared by two techniques. Cross sectional TEM lamellas were extracted by a FIB cutting process. A protective Pt area of 25 x 1.5 x 6 μm is stepwise deposited within the dual beam microscope. To reduce the inflicted damage on the surface of the film, the deposition was conducted in steps of increasing ion beam currents. First a current of 30 pA was used until a thickness of 200 nm was deposited, followed by 0.3 nA for 2 μm and finally 1 nA for another 4 μm. Then, two parallel cross sections were cut adjacent to the Pt-field with dimensions of 25 x 15 x 10 μm and with a Ga^+ current of 7 nA. Subsequently, the lamella is carefully thinned with stepwise decreasing currents depending on the sample and finally removed with a micro manipulator and glued to a TEM grid. In-plane TEM samples were prepared in a conventional routine, similar as proposed by [112]. The Si substrate with the thin film was glued to a glass slide with the Si substrate facing down. The sample was thinned to about 20 μm by grinding and dimpling from the backside. Ar milling was performed to thin the samples until an electron transparent region was obtained. TEM investigations were carried out by Matthias Funk and Byung-Gil Yoo using a Field Emission Gun TEM (Tecnai G^2 F20 X-TWIN, FEI, USA). Micrographs were taken in bright field mode with an accelerating voltage of 200 kV.

The grain sizes of the thin films were measured as an average from line analysis of TEM, FIB channeling contrast and SEM cross section micrographs.

4.2.3. Confocal Light Microscopy

A confocal light microscope (VK-9700, Keyence, Japan) was used to capture larger areas of the thin film samples at lower magnification. The confocal microscope has the advantage that the software allows to examine and analyze micrographs with a large depth of field. Additionally, topographic maps as well as roughness information can be measured. A 150 x objective was used for roughness measurements. The surface roughness at different positions along the cantilever was measured in the same fashion already described for the area analysis (Fig. 4.3 (A)). During image acquisition, the microscope was set to take laser scanning micrographs every 0.01 nm in height and interpolate the resulting intensity to find the correct surface height with a resolution in the nm regime. Afterwards, the microscope program was used to calculate the surface roughness (after the Japanese standard JIS B 0601:1994).

4.2.4. X-Ray Diffraction Analysis

The fatigued thin films were examined in a qualitative manner by synchrotron-based X-Ray Diffraction (XRD) measurements for changes in grain size or orientation. Therefore, the cantilevers were glued onto a Kapton stripe and stepwise scanned along the surface of Cu thin films. The sample surface was oriented perpendicular to the incoming synchrotron beam. XRD scanning experiments and analysis were carried out by Jochen Lohmiller at the High Energy Microdiffraction (HEMD) endstation of beamline ID15A of the ESRF ($E = 69.7$ keV). The setup includes a 165 mm MAR-CCD area detector, which is arranged in transmission to record the diffraction patterns (see Fig. 4.5 (A)). Due to the high energy of the monochromatic synchrotron beam, the cantilever with a thickness of 200 µm, consisting of the Si substrate and the 1 µm Cu thin film, can be easily penetrated. The narrow beam size (8 µm × 20 µm) probes a small volume of the material and therefore enables scanning at differently damaged regions of the thin film. A schematic diffraction pattern is displayed in Fig. 4.5 (B). The radial direction indicates the 2Θ angle and the azimuthal direction Φ captures in-plane orientation. According to Bragg's Law the Debye-Scherrer rings appear at the expected 2Θ positions for an fcc metal ((111), (200), (220), (311), (222)). The diffraction patterns were analyzed with respect to the shape of the rings.

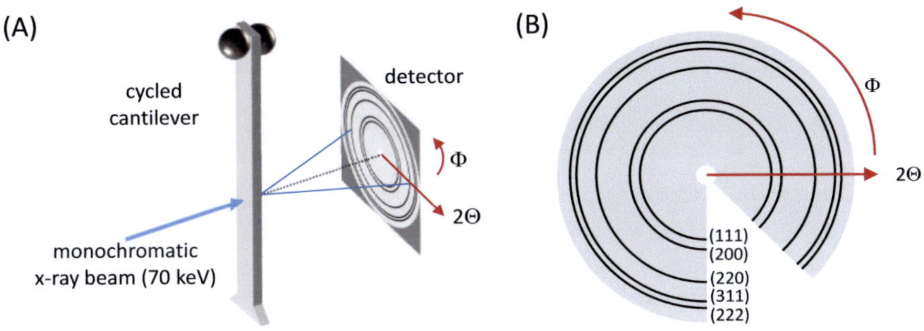

Fig. 4.5: (A) Schematic of the synchrotron XRD setup. (B) Diffraction pattern with the Debye-Scherrer rings at the expected reflexes for an fcc metal.

5. Results of Thin Film Fatigue Testing

In this chapter the results of the thin film fatigue experiments will be presented. Al as well as Cu thin films with different microstructures are the focus of this study. The microstructure was investigated for all tested thin films. Furthermore, a quantitative area analysis and roughness measurements were carried out. Also a qualitative synchrotron analysis is presented. The chapter is divided into two sections for Al and Cu.

5.1. Sample Overview

The samples were fabricated as described in chapter 4.1. The samples, which were successfully tested, are summarized in Table 5.1 and 5.2. The condition was either 'as dep' meaning 'as deposited' with an initially fine-grained microstructure or 'an' meaning 'annealed' for 1 h at 450 °C in a vacuum chamber, where samples possess a coarse-grained microstructure. Cu thin films were additionally deposited on a 10 nm Ti seed layer. Which cantilever was tested, is indicated by the array and cantilever number. Furthermore, the substrate thickness t_{sub} and testing frequencies f_{res} are displayed. The experimental boundary conditions are represented by the given maximum strain amplitude ε_a^{max} and the cycle number.

Table 5.1: Sample overview of successfully tested Al thin films.

material	condition	array number	cantilever number	t_{sub} (μm)	f_{res} (Hz)	ε_a^{max} (%)	cycle number	
Al	'as dep'	1	2	196	651	0.180	5×10^7	
Al	'as dep'	1	3	192	620	0.112	1×10^9	
Al	'as dep'	1	4	189	597	0.180	1×10^8	
Al	'as dep'	1	5	187	570	0.179	1×10^8	**
Al	'as dep'	1	7	184	534	0.198	5×10^6	
Al	'as dep'	3	6	182	630	0.146	5×10^7	*
Al	'as dep'	3	7	182	611	0.150	1×10^8	*
Al	'as dep'	3	9	190	608	0.157	2×10^8	*
Al	'an'	2	11	199	652	0.194	5×10^6	
Al	'an'	2	12	199	639	0.192	2×10^7	***
Al	'an'	2	13	196	615	0.138	2×10^8	
Al	'an'	2	15	200	600	0.209	2×10^7	*
Al	'an'	2	17	197	580	0.166	1×10^8	
Al	'an'	2	18	197	573	0.161	5×10^7	
Al	'an'	2	19	197	558	0.122	2×10^7	

*Comments: 'as dep' = 'as deposited' thin films; 'an' = 'annealed' thin films; * illumination problem; ** experiment was interrupted; *** data acquisition problem; **** small surface defects on the thin film.*

Table 5.2: Sample overview of successfully tested Cu thin films.

material	condition	array number	cantilever number	t_{sub} (µm)	f_{res} (Hz)	ε_a^{max} (%)	cycle number	
Cu	'as dep'	1	1	191	646	0.187	1×10^8	
Cu	'as dep'	1	2	190	615	0.211	5×10^7	***
Cu	'as dep'	1	4	187	565	0.216	1×10^8	
Cu	'as dep'	1	5	186	555	0.206	2×10^7	
Cu	'as dep'	1	6	183	544	0.173	2×10^8	
Cu	'as dep'	1	7	184	523	0.174	1×10^8	****
Cu	'as dep'	1	8	184	516	0.195	1×10^7	***
Cu	'an'	3	1	183	607	0.192	2×10^7	
Cu	'an'	3	4	181	546	0.132	1×10^9	
Cu	'an'	3	5	186	542	0.146	1×10^8	
Cu	'an'	3	6	199	568	0.179	5×10^7	
Cu	'an'	3	7	196	559	0.192	5×10^7	
Cu	'an'	3	9	194	622	0.212	3×10^8	
Cu	'an'	3	11	207	632	0.180	2×10^7	
Cu	'an'	3	12	209	616	0.179	2×10^8	
Cu 'on Ti'	'as dep'	1	2	182	426	0.141	4×10^7	*
Cu 'on Ti'	'as dep'	1	3	182	374	0.163	2×10^7	*
Cu 'on Ti'	'as dep'	1	5	182	825	0.150	2×10^7	*
Cu 'on Ti'	'as dep'	1	6	188	628	0.177	5×10^7	****
Cu 'on Ti'	'as dep'	1	8	185	585	0.174	2×10^8	****

*Comments: 'as dep' = 'as deposited' thin films; 'an' = 'annealed' thin films; * illumination problem; ** experiment was interrupted; *** data acquisition problem; **** small surface defects on the thin film.*

5.2. Al Thin Films

The Al thin films have a thickness of around 1 µm and a fine- or coarse-grained microstructure. The thin film samples were cycled and the damage evolution was investigated.

5.2.1. Microstructural Investigation of Al Thin Films

In this paragraph the microstructural investigations of Al thin films before and after cycling are presented. The different fatigue damage features will be introduced and the overall damage evolution is shown.

Initial Microstructure

Fig. 5.1 shows representative SEM top view micrographs of the 'as deposited' and 'annealed' Al thin films before cycling. The grain sizes *d* of the 'as deposited' thin films (Fig. 5.1 (A)) are in the range of 800 nm with a partially globular shape. The top surface shows surface features like grooving and single hillocks. The 'annealed' thin films (Fig. 5.1 (B)) have columnar grains with a size of about 1200 nm. The surface in this case shows pore formation and a more pronounced hillock formation.

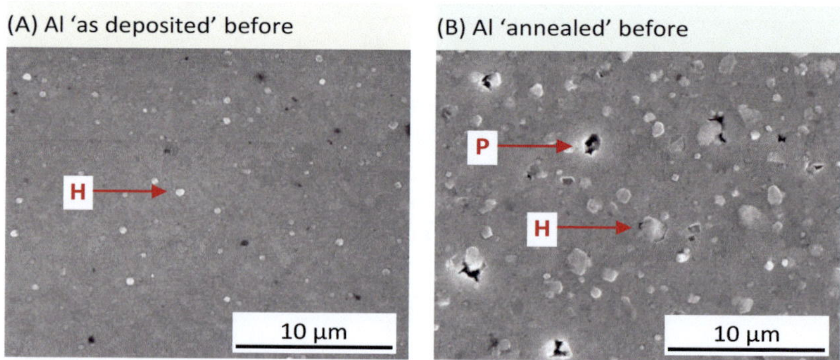

Fig. 5.1: SEM micrographs of (A) 'as deposited' and (B) 'annealed' Al thin films before fatigue testing. H = hillock, P = pore.

Fatigue Damage Features

In the following paragraphs a more detailed examination of the fatigue damage features is performed in the highly damaged regions of Al thin films by high magnification top view SEM micrographs and FIB milled cross sections.

Fig. 5.2 displays an 'as deposited' Al thin film after 10^8 cycles with a local strain amplitude of $\varepsilon_a = 0.18$ % in top view (A), as well as a cross section (B). The surface shows areas of non uniformly shaped hillocks, where whole grains are pushed out of the thin film. Additionally, areas with coarse extrusions with a lamella size of around 200 nm can be identified. Within a certain area the extrusion lamellas are oriented parallel to each other. The overall orientation of the extrusions is more or less 45 ° to the bending direction. Formation of short cracks is less pronounced but can be seen along grain boundaries. The cross section reveals pore formation at the substrate interface and throughout the cross section. Grain growth can be observed. The grains after fatigue are in the range 1200 nm. The film thickness changes locally and is not uniform anymore.

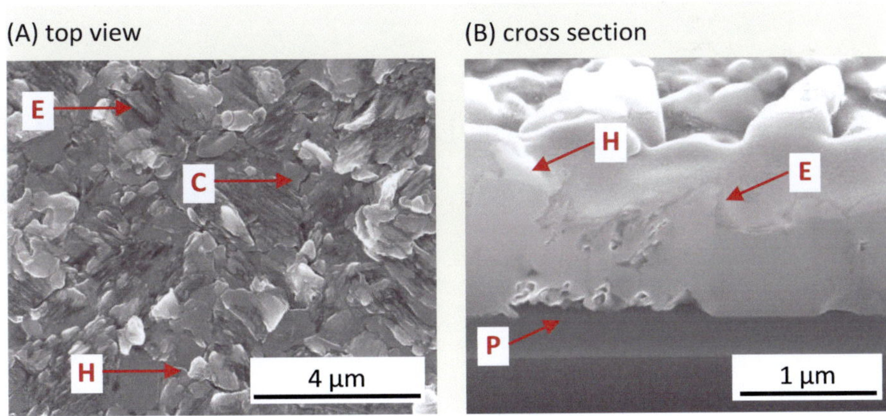

Fig. 5.2: SEM micrographs of an Al thin film 'as deposited' after 10^8 cycles with a local strain amplitude of $\varepsilon_a = 0.18$ %. (A) top view and (B) cross section. H = hillock, P = pore, E = extrusion, C = crack.

The damage structures in top view (A) and cross section (B) SEM micrographs of an 'annealed' Al thin film are shown in Fig. 5.3. The film was fatigued to 10^8

cycles with a local strain amplitude of $\varepsilon_a = 0.17$ %. The damage structure differs slightly from the 'as deposited' condition. Similarly, non uniformly shaped hillocks can be identified. The hillocks formed during annealing and cycling cannot be differentiated. The extrusion areas show finer extrusion lamellas compared to the 'as deposited' condition with a size of around 50 nm. In some cases the extrusion lamellas are oriented parallel within an extruded region, but the extrusion can be randomly oriented as well. In this case the extrusion lamellas do not span over the whole region but are shorter and protrude from the surface. Short cracks are formed between the extrusion areas. In the cross section micrographs no explicit grain growth has been identified. Pore formation occurred mainly within the film interior and less at the interface. The film thickness varies locally and is not uniform.

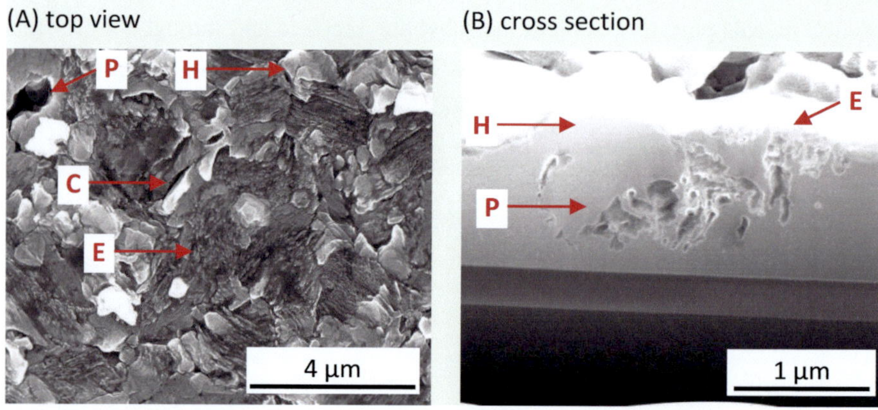

Fig. 5.3: SEM micrographs of an 'annealed' Al thin film after 10^8 cycles with a local strain amplitude of $\varepsilon_a = 0.17$ %. (A) top view and (B) cross section. H = hillock, P = pore, E = extrusion, C = crack.

Overall Damage Evolution

The overall damage evolution on the top surfaces after 10^8 cycles of an 'as deposited' and 'annealed' Al thin films are displayed in SEM micrographs in Fig. 5.4 (A) and (B) respectively. The micrographs were taken in the center along the cantilevers as described in chapter 4.2.1. This means that every micrograph corresponds to a specific strain amplitude which increases from top to bottom. In both microstructural conditions damage at the surface can be observed to be a mixture of hillock, extrusion, and crack formation. The amount of fatigue damage increases with increasing strain amplitude. Differences in damage formation depending on the initial microstructure can be observed.

Going from low to high strain amplitudes, the 'as deposited' Al thin films (Fig. 5.4 (A)) show roughening of the surface at $\varepsilon_a = 0.10$ % caused by hillock formation. At a strain amplitude of $\varepsilon_a = 0.12$ % the amount of hillocks has increased in number and size. A further increased amplitude of $\varepsilon_a = 0.14$ % additionally displays first extrusion formation. At an amplitude of $\varepsilon_a = 0.16$ % the areas with extrusions have grown whereas the amount of hillocks stays the same. Furthermore, some short cracks between damaged areas can be observed. In the case of an amplitude of $\varepsilon_a = 0.18$ % the extrusion areas have grown to their final size and a homogeneously damaged surface has resulted.

The 'annealed' Al thin films (Fig. 5.4 (B)) display a similar damage structure. For a strain amplitude of $\varepsilon_a = 0.09$ % fatigue damage cannot be identified by comparing it to the initial microstructure, as the hillocks which can be seen were already caused by annealing. If the amplitude is $\varepsilon_a = 0.11$ % the number of hillocks has increased and for $\varepsilon_a = 0.13$ % additional regions with extrusions have formed. The pores formed by annealing are still visible. A promoted extrusion formation can be recognized for a further increase in strain amplitude at $\varepsilon_a = 0.15$ % and the annealing pores seem to close. Instead, cracking can be found between damaged areas. In the case of an amplitude of $\varepsilon_a = 0.17$ %, the damage structure has fully developed, where the extruded areas have grown to their final size. The amount of hillocks has not further increased. Similar to the 'as deposited' film, a homogeneously damaged surface is seen in the end. A quantitative analysis of the damage formation is part of chapter 5.2.2.

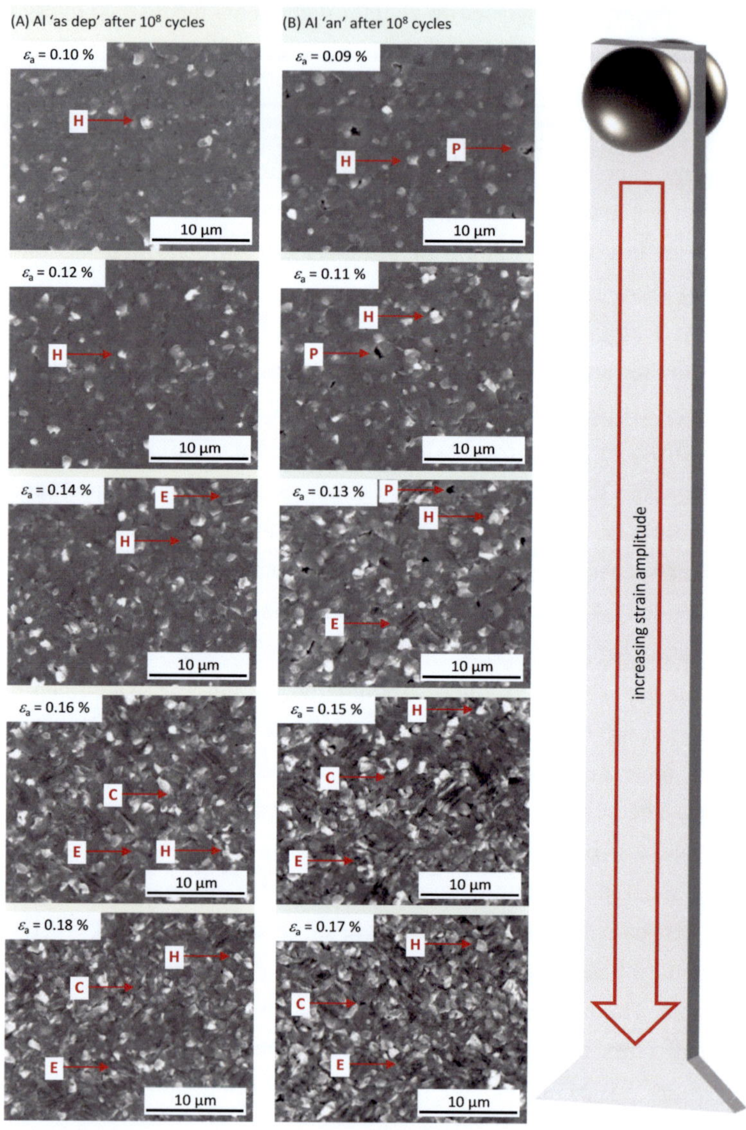

Fig. 5.4: SEM top view micrographs of Al thin films (A) 'as deposited' and (B) 'annealed' after 10^8 cycles. Micrographs were taken at different positions along the cantilever, which correspond to different strain amplitudes. Different damage features at the film surface are indicated: H = hillock, P = pore, E = extrusion, C = crack.

5.2.2. Quantitative Investigation of Al Thin Films

The surface quality of 'as deposited' and 'annealed' Al thin films after cycling is investigated by area and line analyses and roughness measurements. Fig. 5.5 displays the evolution of the relative intensity, the normalized measured undamaged area and the normalized surface roughness with their standard deviation along the position *x* of the Al cantilevers. For both Al conditions the reflectivity displays a minimum in the high strained region and increases towards the free end. The measurement of the fraction of the undamaged area is in good agreement with the reflectivity. There is also a minimum in the high strained region, and the amount of undamaged area increases toward the free. The curves show the same trend within the accuracy of measurement. The scatter of the area analysis is in some cases quite high due to uncertainties in binarizing the micrographs. Another indication that the reflectivity of the laser gives reasonable data is proven by the surface roughness measurement. In both cases the surface roughness shows a maximum which corresponds to the minimum of reflectivity. The roughness decreases towards the free end of the cantilever which indicates less damage.

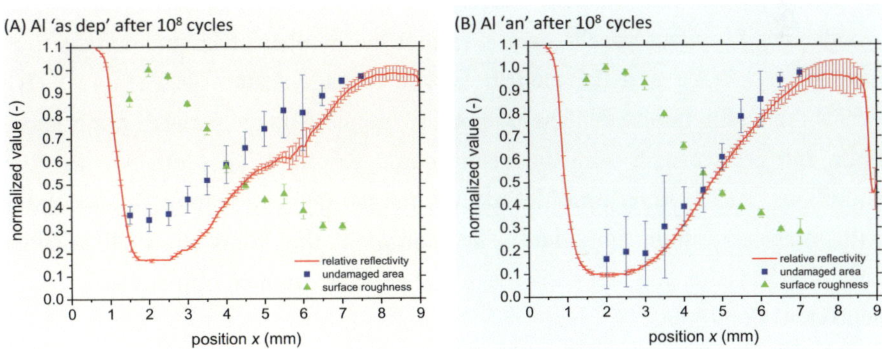

Fig. 5.5: Quantitative analysis of (A) 'as deposited' and (B) 'annealed' Al thin films after fatigue testing. For comparison, the relative reflectivity, the normalized undamaged area and roughness are plotted versus the position x along the cantilever.

By using a line analysis the size of individual fatigue damage features can be measured and is plotted as a cumulative distribution function in Fig. 5.6 for the 'as deposited' (A) and 'annealed' (B) condition. It was not distinguished between the individual damage features, hillocks and extrusion areas. For both thin films it can be seen that with increasing strain amplitude the distribution is shifted to larger feature sizes. Especially, the probability of finding larger fatigue damage features is more pronounced for the higher strain amplitudes.

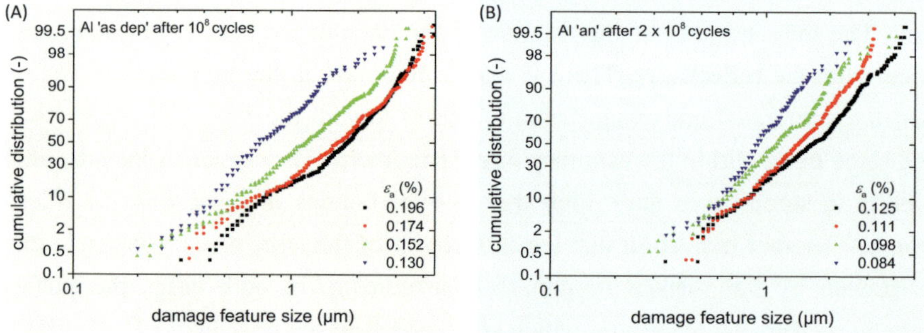

Fig. 5.6: Cumulative distribution of the damage feature size of (A) 'as deposited' and (B) 'annealed' Al thin films for different strain amplitudes.

Another way to visualize the data is to plot the number of fatigue damage features (Fig. 5.7 (A)) and the median fatigue damage feature size (Fig. 5.7 (B)) versus the strain amplitude. The number of fatigue damage features has a maximum and are constant with increasing strain amplitude for both Al thin film conditions. At the same time the median fatigue damage feature size increases with increasing strain amplitude. The 'annealed' thin films show all together larger fatigue damaged areas and reach it at smaller strain amplitudes than the 'as deposited' thin films.

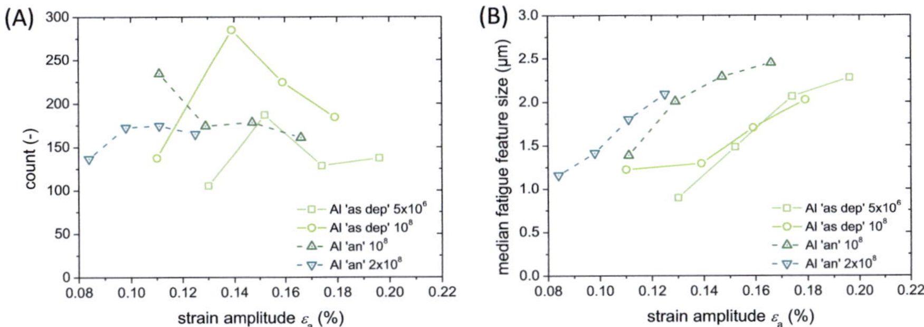

Fig. 5.7: (A) Comparison of number of fatigue damage features and (B) the median fatigue feature size for Al thin films 'as deposited' and 'annealed' after cycling.

5.2.3. Lifetime Diagrams of Al Thin Films

With the method described in chapter 3.3.3 the lifetime data for the tested cantilevers has been obtained. A summary of all the successfully tested 'as deposited' and 'annealed' Al thin films is shown in Fig. 5.8 for a lifetime criterion of relative reflectivity is 0.8. In a typical S-N curve the strain amplitude is plotted versus the number of cycles to failure both on logarithmic axes. The grey symbols represent the samples where difficulties occurred during testing (see Table 5.1) and are not taken into account for further analysis. The error bars are obtained from the error propagation in chapter 3.4.2 and will not be taken into account in detail for further plotting as the data in both cases lies within an error of 10 % marked by the colored area. With decreasing strain amplitude the number of cycles to failure increases. The 'as deposited' and 'annealed' Al thin films follow the Basquin Law (Eq. 2.2).

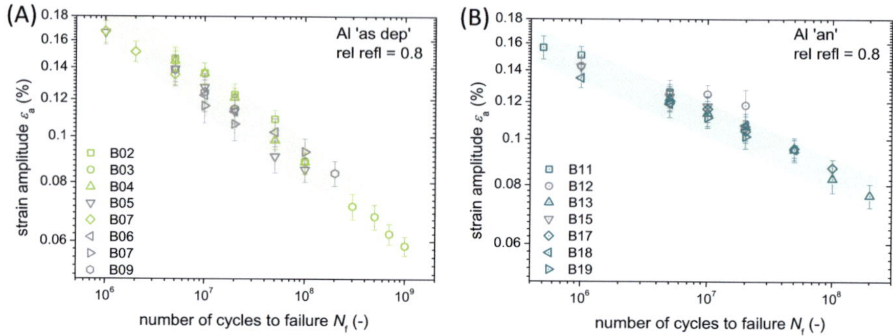

Fig. 5.8: Lifetime diagrams for successfully tested (A) 'as deposited' and (B) 'annealed' Al thin films for a failure criterion of a relative reflectivity of 0.8. The grey symbols indicate uncertainties in the measurement.

For a better comparison, the data for the trusted Al cantilevers is plotted in Fig. 5.9. It can be seen that the fine-grained 'as deposited' Al thin films display higher lifetimes than the coarse-grained 'annealed' thin films. However, for smaller amplitudes the lifetimes coincide and the sensitivity with respect to the strain amplitude seems to increase. The data was fitted to obtain the parameters of the Basquin Law (Eq. 2.2) and are summarized in Table 5.3. Furthermore, the data shows a steady decrease in lifetime and an endurance limit cannot be observed in the investigated regime.

In Fig. 5.10 selected cantilevers for both conditions are plotted with different failure criteria ranging from a relative reflectivity of 0.9 as conservative criterion, up to 0.5 as non conservative. In general, the overall trend is maintained and the curves are shifted towards higher lifetimes for decreasing reflectivities. Moreover, a trend is seen that the sensitivity factor decrease with increasing damage criterion, particularly for the 'as deposited' thin films.

Table 5.3: Fatigue data of Al thin films.

sample	fatigue sensitivity factor b	fatigue strength σ_f (MPa)
Al 'as dep'	-0.14	913
Al 'an'	-0.11	467

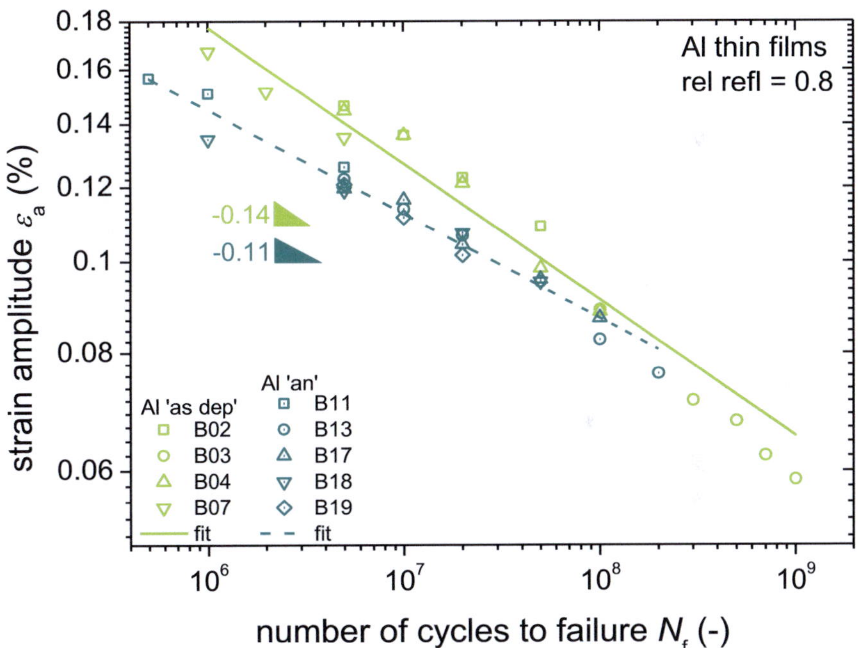

Fig. 5.9: Lifetime diagram of 'as deposited' and 'annealed' Al thin films with a damage criterion of relative reflectivity of 0.8.

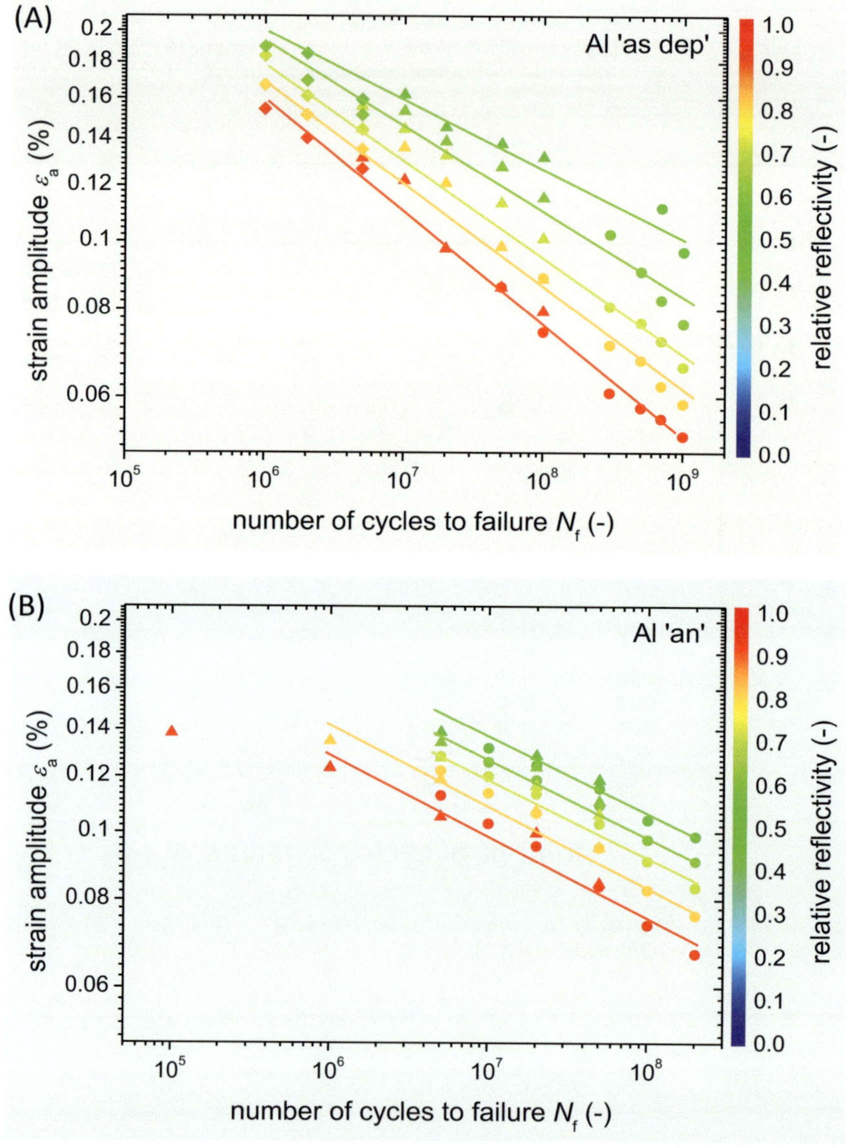

Fig. 5.10: Lifetime diagrams of (A) 'as deposited' and (B) 'annealed' Al thin films with different failure criteria.

5.3. Cu Thin Films

In this chapter the results for the 1 μm Cu thin films are presented. There are three different types of Cu thin films. Fine-grained 'as deposited' Cu, coarse-grained 'annealed' Cu, and Cu 'on Ti' seed layer. First, the damaged thin films are characterized qualitatively by microstructural investigation and X-Ray Diffraction analysis. Additionally, the thin films were analyzed quantitatively by line and area analysis. In the end, the lifetime data obtained from the experiments are shown and described.

5.3.1. Microstructural Investigation of Cu Thin Films

In this paragraph the microstructural investigations of Cu thin films before and after cycling are presented. The different fatigue damage features will be introduced and the overall damage evolution is shown.

Initial Microstructure

In Fig. 5.11 a comparison of all the Cu thin films ('as deposited' (A), 'annealed' (B), 'on Ti' (C)) before testing is demonstrated. For each sample a micrograph has been taken by SEM, FIB, and TEM. The 'as deposited' Cu thin films have globular grains with a size of 200 nm. The surface displays circular shaped features with grooves. The FIB micrograph reveals differently oriented small grains from channeling contrast, as well as some extremely large initial grains. In the in-plane TEM micrograph a fine-grained microstructure with some twinning can be identified. The 'annealed' Cu thin films display surface pores in the SEM micrograph. The FIB micrograph reveals a coarser grained microstructure with different orientation by the channeling contrast. The grains have a size of about 1200 nm and are columnar. Twins were observed in the TEM micrograph. The surface of the Cu deposited 'on Ti' displays grooves and a facetted structure. The grains are on the order of 400 nm with globular shape. The channeling contrast of the FIB micrograph reveals the fine-grained structure with a large amount of similarly oriented grains. The TEM micrograph verifies the fine-grained microstructure and less twinning.

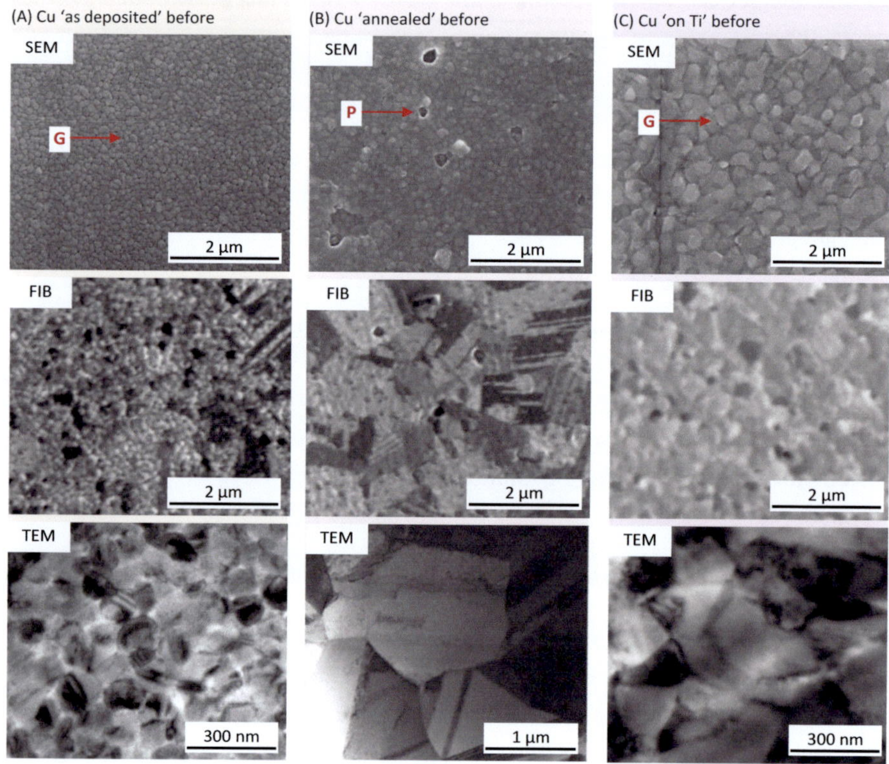

Fig. 5.11: SEM, FIB and TEM micrographs of (A) 'as deposited', (B) 'annealed' and (C) 'on Ti' Cu thin films before cycling. G = grooves, P = pore.

Fatigue Damage Features

The following paragraphs focus on the fatigue damage feature formation in Cu thin films by high magnification top view SEM micrographs, FIB milled cross sections and plane view TEM micrographs.

The damage formation of an 'as deposited' Cu thin film with a strain amplitude of $\varepsilon_a = 0.22$ % after 10^8 cycles is shown in Fig. 5.12. The top view SEM micrograph in (A) displays extrusion islands. The extrusions lamellas are mostly oriented with an angle of 45° relative to the beam axis. Undamaged regions can be easily identified, as the original surface structure is still preserved. The cross section in (B) reveals additional damage throughout the film thickness. Pores form within the cross section as well as at the interface of Cu and Si. The pores are lined up in parallel lines which also feature a 45 ° angle. Although damage occurred, the film thickness stayed at about 1 μm but is folded, which is correlated to the extrusions at the surface and the underneath formed pores. The grains have grown to a size of over a 1 μm which is in the range of the film thickness. The in plane TEM micrograph in (C) shows coarsened grains with twins, but the identification of dislocations is difficult.

The fatigued 'annealed' Cu thin film and its damage structures for a strain amplitude of $\varepsilon_a = 0.18$ % are shown in Fig. 5.13. In the SEM top view micrograph (A), the damaged and undamaged areas can be distinguished by the extrusion formation. The undamaged areas show the original surface structure. The extrusion lamellas are randomly oriented but are parallel within an individual extrusion island. The lamellas have different spacing, either they are rather coarse with 500 nm or fine within a region and a spacing of less than 100 nm. The pores at the surface formed during annealing act as short crack initiation sites at the surface or are incorporated into the extrusion islands. Additionally, twins have an influence on damage formation, as a twin boundary can be identified by a straight boarder of an extrusion island. The cross section in (B) reveals the damage formation throughout the film thickness. Pores form near the surface, within the film thickness and at the interface. The alignment of the pores along a 45 ° line is less pronounced but can be still recognized. The same applies for the folding of the Cu thin films according to extrusions and pores at the interface. The TEM plane view micrograph in (C) shows some tangled dislocation within the grains, but the formation seems to be predominantly at and inside the twins. No ordered dislocation structure like walls and cells, which occur in bulk material, are seen.

The damage morphology of a Cu thin film 'on Ti' seed layer after 2 x 10^8 cycles and a strain amplitude of $\varepsilon_a = 0.17$ % is shown in Fig. 5.14. The damage structure in this case differs from the previously shown Cu thin films. The SEM top view micrograph in (A) shows a detail of an extrusion island of the Cu 'on Ti' thin film. Similarly as before, damaged and undamaged regions can be identified, whereas the undamaged regions maintain their original surface structure. The extrusion islands show coarse and ultra fine extrusion lamellas. The coarse extrusion lamellas are partially parallel oriented and have an irregular shape. In between, there are ultra fine extrusions which protrude from the surface. There is no favored orientation of these extrusions. Short surface cracks predominantly occur between damaged and undamaged regions, but also in between differently oriented extrusion lamellas. The cross section reveals damage throughout the thin film but different than before. Pores form only near the surface of the Cu thin film and there is no pore formation at the interface. The pores near the surface line up in two perpendicular 45 ° oriented lines. The pores are smaller compared to the other Cu thin films and are more closely packed. As the Cu thin film is still attached to the Ti seed layer, the Cu thin film cannot be folded but rather has an irregular thickness. The grains have grown to a size of over 1 µm and can be observed in the TEM micrograph in (C). Multiple dislocations can be identified within several grains. They do not form high ordered structures, but they can be called tangled dislocations. Furthermore, no twins were observed.

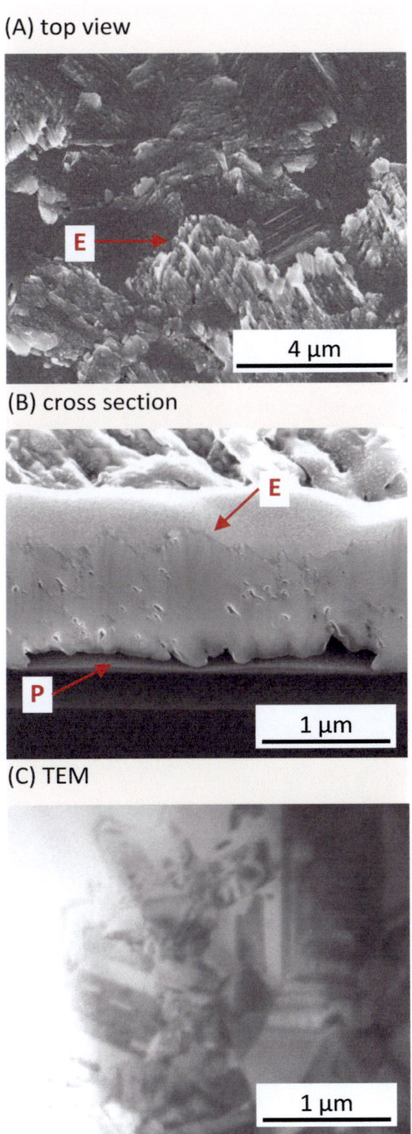

Fig. 5.12: Micrographs of an 'as deposited' Cu thin film after 10^8 cycles with a strain amplitude of $\varepsilon_a = 0.22$ %. (A) SEM top view, (B) SEM cross section, (C) plane view TEM. E = extrusion, P = pore.

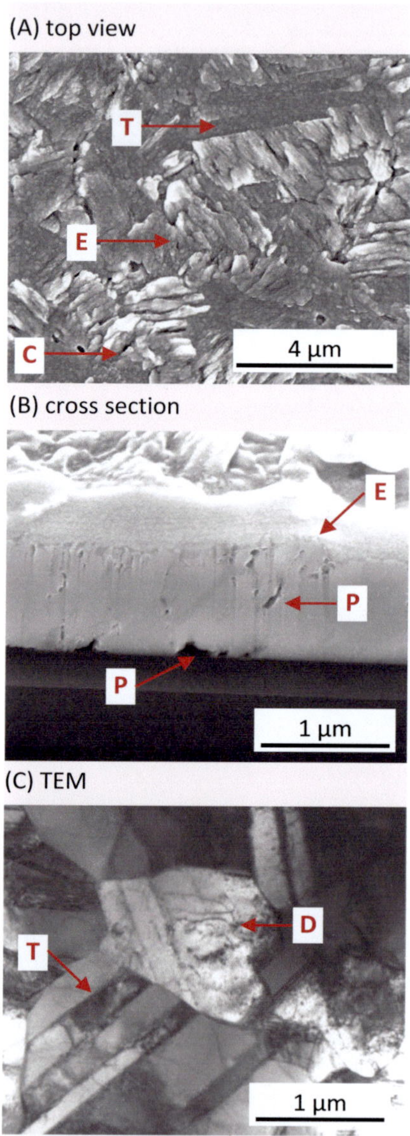

Fig. 5.13: Micrographs of an 'annealed' Cu thin film after 2 x 10^8 cycles at a strain amplitude of $\varepsilon_a = 0.18$ %. (A) SEM top view, (B) SEM cross section, (C) plane view TEM. E = extrusion, P = pore, C = crack, T = twin, D = dislocation.

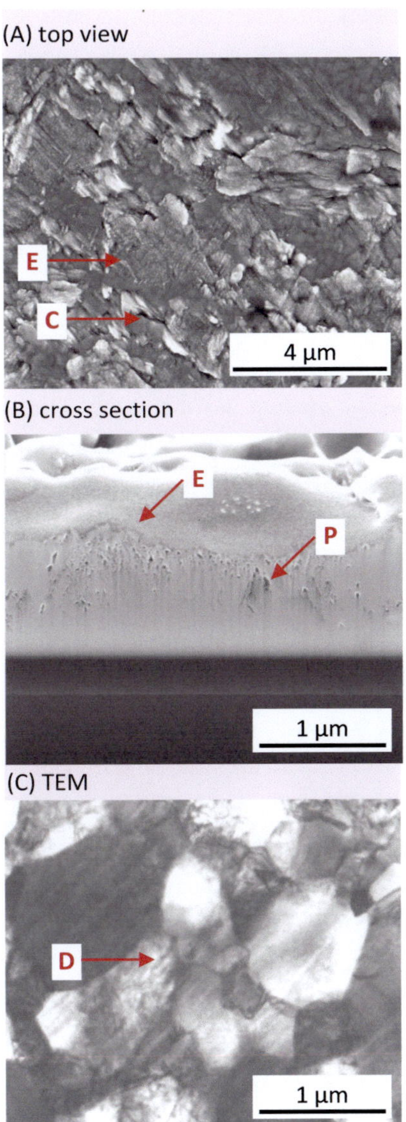

Fig. 5.14: Micrographs of a Cu thin film 'on Ti' seed layer after 2 x 10^8 cycles with a strain amplitude of $\varepsilon_a = 0.17$ %. (A) SEM top view, (B) SEM cross section, (C) plane view TEM. E = extrusion, P = pore, C = crack, D = dislocation

Overall Damage Evolution

The surface morphology of the three Cu thin films after cycling is displayed in Fig. 5.15 for 'as deposited' (A), 'annealed' (B), and 'on Ti' (C) Cu thin films. The cycle number and strain amplitude are indicated in the micrograph. Generally, with increasing strain amplitude the amount of damage increases, too.

After 10^8 cycles, going from low to high strain amplitudes, the 'as deposited' Cu thin film (Fig. 5.15 (A)) shows first damage at a strain amplitude of $\varepsilon_a = 0.15$ %. First extrusions form with a lamella spacing of 300 nm at various positions throughout the thin film. At an amplitude of $\varepsilon_a = 0.17$ % the extruded islands have grown and the amount has increased. The extrusion lamellas have no specific orientation. By further increasing the amplitude to $\varepsilon_a = 0.19$ %, the size and amount of the extrusion islands has increased. For the maximum amplitude of $\varepsilon_a = 0.22$ % the extrusion islands have grown further to a size of about 3 µm.

The situation differs for the 'annealed' Cu thin films (Fig. 5.15 (B)). After 2 x 10^8 cycles, damage formation starts at $\varepsilon_a = 0.12$ % with small extrusions forming predominantly near the annealing pores. For $\varepsilon_a = 0.14$ % the extrusion islands have grown and new sites of extrusion islands have been nucleated. A further increase in extrusion island size and amount can be identified for a strain amplitude of $\varepsilon_a = 0.16$ %. The extrusion lamellas are parallel and have a variable spacing size but do not show a specific orientation. In between the extrusion lamellas and between the extrusion islands short cracks can be observed. At the maximum strain amplitude of $\varepsilon_a = 0.18$ % the extrusion islands have further densified with a median size of 3 µm.

In the case of the Cu thin films 'on Ti', the damage evolution takes place in the following way. For an amplitude of $\varepsilon_a = 0.12$ % no damage could be observed after 2 x 10^8 cycles. The next micrograph taken at a strain amplitude of $\varepsilon_a = 0.14$ % already shows large extrusion islands. The extrusion lamella size is on the order of 300 nm, but the lamella itself appears to have a more irregular shape than in the other cases. The different islands have already grown together. A way to distinguish between them is the change in orientation of the extrusion lamellas. Short cracks can be identified between damaged and undamaged regions or along the extrusions lamellas. At a strain amplitude of $\varepsilon_a = 0.16$ %, the amount of extrusion islands has increased. Extrusion lamellas with irregular unparallel lamellas can be observed. At a maximum amplitude of $\varepsilon_a = 0.17$ % the fraction of damaged area has further increased. The lamellas within an island seem to have lost their parallelism compared to lower strain amplitudes. The lamella spacing is difficult to measure and compared to the other Cu thin film types. Qualitatively, they appear to be finer and more irregular.

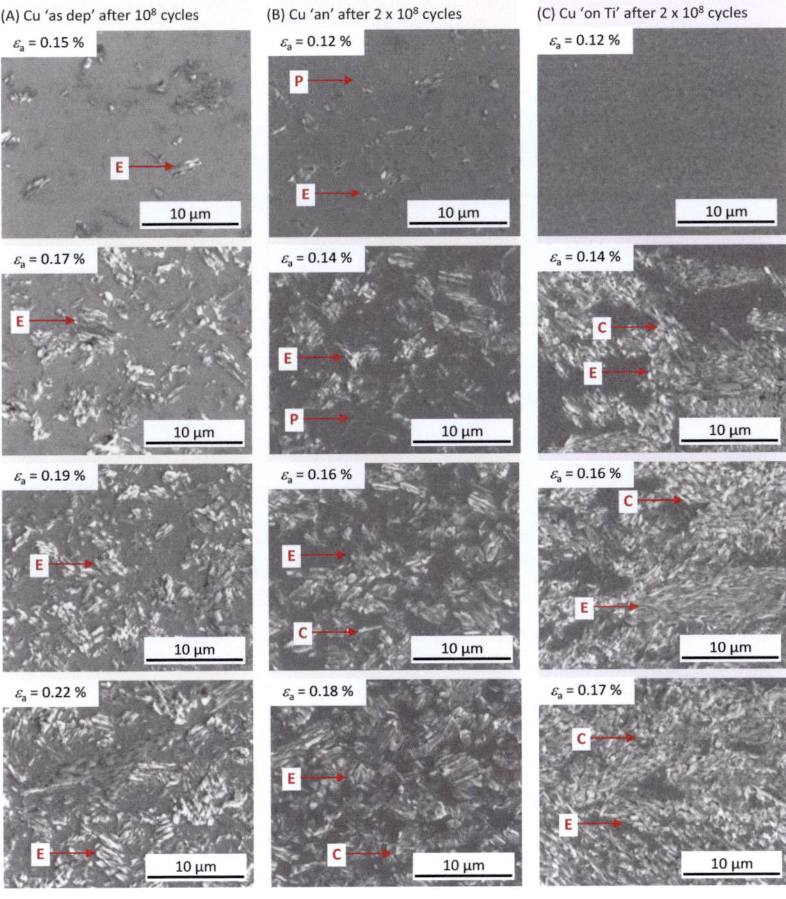

Fig. 5.15: SEM top view micrographs of cycled Cu thin films (A) 'as deposited', (B) 'annealed' and (C) 'on Ti' at different strain amplitudes. E = extrusion, P = pore, C = crack.

5.3.2. X-Ray Diffraction Analysis of Cu Thin Films

Another way to qualitatively investigate the microstructure of the Cu thin films was conducted by synchrotron based X-Ray Diffraction (XRD). To better understand the evolution of the damage, all types of Cu thin films were measured with a synchrotron beam (see 4.2.4). Quantitative analysis, e.g. line broadening, is not possible due to poor statistics resulting from relatively large grains compared to the narrow beam size. Instead, the scan along the cantilever of the cycled thin film is used as proof of concept to identify the damaged areas and the changes in microstructure by means of locally recorded diffraction patterns. Fig. 5.16 displays the diffraction patterns for Cu thin films after cycling for (A) 'as deposited', (B) 'annealed', and (C) Cu 'on Ti' seed layer. The shallow reflections, which can be seen in all diffraction patterns in the background, result from the (100) Si substrate and are neglected.

The diffraction patterns in (A) for the 'as deposited' thin film first display almost complete Debye-Scherrer rings in the undamaged area, which indicate a fine-grained microstructure. The diffraction rings appear at the typical radii for an fcc material. Several spots with higher intensity are an indication of individual large grains. This is in agreement with the microscopic investigation, where several large grains were observed (Fig. 5.11 (A)). The Debye-Scherrer rings dissolve with increasing strain amplitudes to single spots. The dissolution of the full ring into single spots is typically attributed to grain growth, as individual large grains yield high intensities at a certain combination of 2 Θ and Φ based on the grain orientation. The spots remain on the position of the original ring but do not show a specific in-plane orientation.

In the case of the 'annealed' Cu thin film the diffraction patterns in (B) display hot spots superimposed to the ring contours for all of the diffraction patterns. The damage cannot be identified easily in this case, as the grains are large in the undamaged and highly damaged region. The grains display no specific in-plane orientation.

The diffraction patterns for the Cu 'on Ti' thin film in (C) show partial Debye-Scherrer rings in the undamaged region. This may be an indication for a texture that results from the growth of the film on the Ti interlayer. Again, the continuous signal in the undamaged conditions changes to single spots at different Φ-angles on the contour of the ring after 2 x 10^7 cycles.

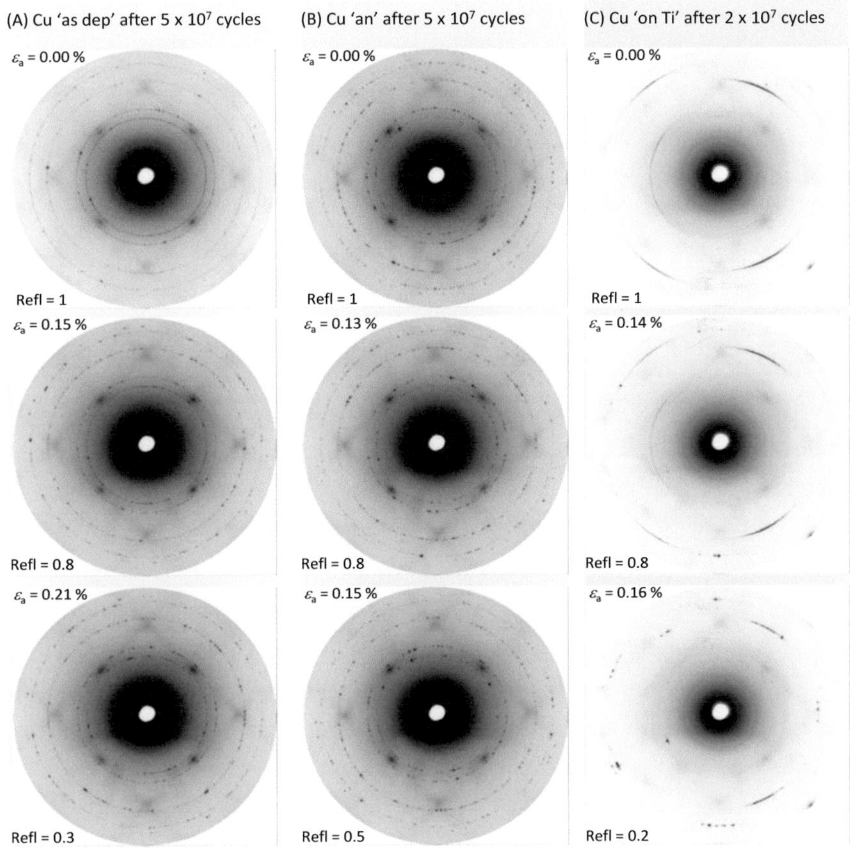

Fig. 5.16: Diffraction patterns at different strain amplitudes for (A) 'as deposited', (B) 'annealed', (C) 'on Ti' Cu thin films.

5.3.3. Quantitative Investigation of Cu Thin Films

The quantitative comparison of the reflectivity and the area analysis of the undamaged area is plotted in Fig. 5.17 for an 'as deposited' (A), 'annealed' (B), and 'on Ti' (C) Cu thin film after cyclic loading. Measuring the roughness of the

Cu thin film described in chapter 4.2.3 does not work in this case, as the extrusions are too fine and errors occur due to unmeasureable angles. The trend for all three samples is similar to the results of the Al thin film case. The amount of undamaged area has a minimum which corresponds to the minimum in reflectivity. With increasing position *x* along the cantilever the intensity increases as well as the undamaged area. The agreement is fairly good for the 'annealed' and 'on Ti' Cu thin film and less for the 'as deposited' thin film, but the trend is the same.

The cumulative distribution functions of the extrusion island sizes are demonstrated in Fig. 5.18 for (A) 'as deposited', (B) 'annealed', and (C) 'on Ti' Cu thin film after cycling at different strain amplitudes. For the 'as deposited' and 'annealed' Cu thin films, the distribution function is shifted to increasing extrusion island sizes with increasing strain amplitude. The cumulative distribution function for the Cu thin film 'on Ti' is constant in the median region for the different strain amplitudes. There are only slight variations in the smallest and largest island sizes.

To find out if the extrusion islands grow or if they are just getting more numerous, the amount and the median extrusion island size is analyzed and plotted versus the strain amplitude for the different Cu thin films in Fig. 5.19. The number of extrusion islands in the case of the 'as deposited' Cu thin film increases with increasing strain amplitude until saturation or even decreases. The size of the extrusion islands increases steadily with increasing strain amplitude. The situation is similar for the 'annealed' Cu thin films. The amount of extrusion islands increases with increasing strain amplitude until saturation. The amount of fatigue damage areas is larger in the 'annealed' Cu thin film, than in the 'as deposited' thin film. The size of the extrusion islands increases simultaneously with increasing strain amplitude and is similar to the 'as deposited' Cu thin films. In the case of the Cu thin films 'on Ti', the amount of extrusion islands increases with increasing strain amplitude and is lower than for the other thin films. The initiated extrusion islands have a size of 4 μm from the beginning and do not grow significantly with increasing strain amplitude. The extrusion island size is also larger than the other Cu thin films.

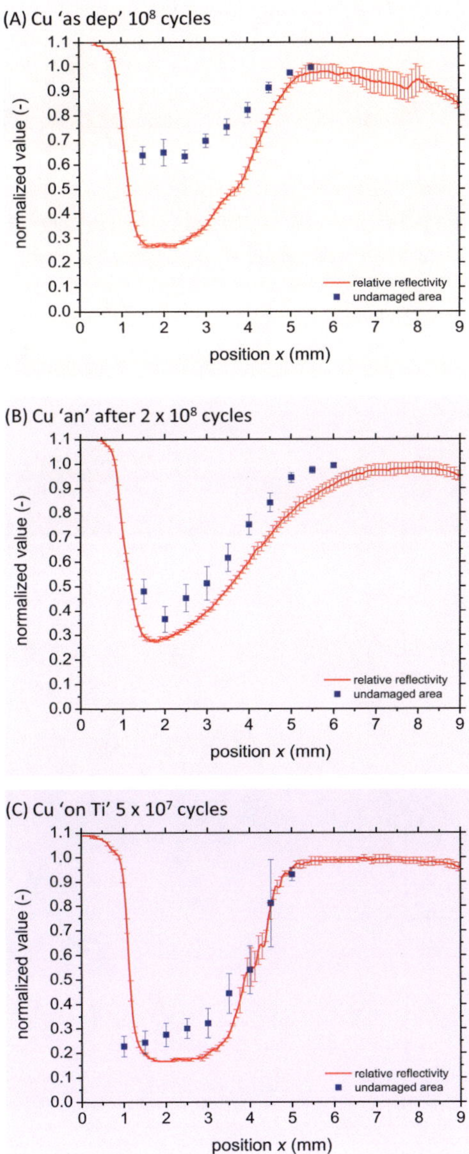

Fig. 5.17: Comparison of area analysis and the measured reflectivity along the center of the cantilever for (A) 'as deposited', (B) 'annealed', (C) Cu 'on Ti' Cu thin films.

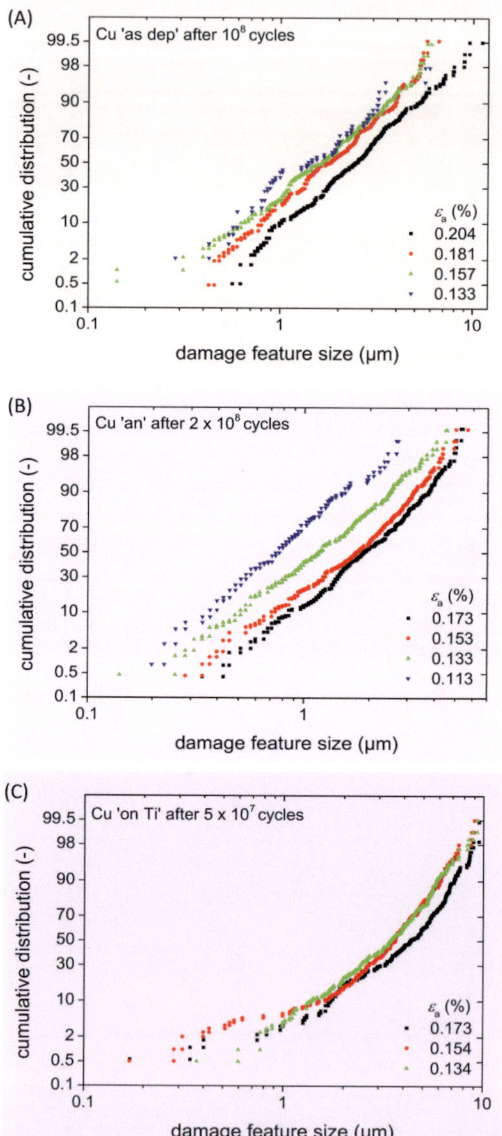

Fig. 5.18: Cumulative distribution of the damage feature size of (A) 'as deposited' (B) 'annealed' and (C) 'on Ti' Cu thin films for different strain amplitudes.

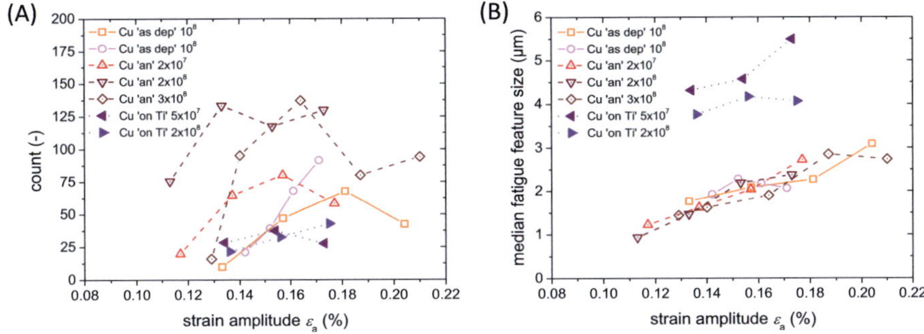

Fig. 5.19: Comparison of (A) number of fatigue damage size and (B) the median fatigue damage size for Cu thin films 'as deposited', 'annealed', and 'on Ti' seed layer.

5.3.4. Lifetime Diagrams of Cu Thin Films

In this chapter the lifetime diagrams obtained from the reflectivity scans (see 3.3.3) for the three different types of Cu thin films are presented. Fig. 5.20 displays all successfully tested cantilevers for the Cu thin films, whereas the grey symbols indicate some uncertainties in the measurement. The strain amplitude is plotted versus the number of cycles to failure in logarithmic scaling. The damage criterion for all the plots is a relative reflectivity of 0.8. The error bars are obtained from the error analysis described in chapter 3.4.2. All the diagrams have in common that with decreasing strain amplitude the lifetime increases. The data of the individual cantilevers lies within the error bars. For the 'annealed' Cu thin film in (B) the lifetime reaches a plateau at a strain amplitude of around $\varepsilon_a = 0.11$ %, which indicates an endurance limit.

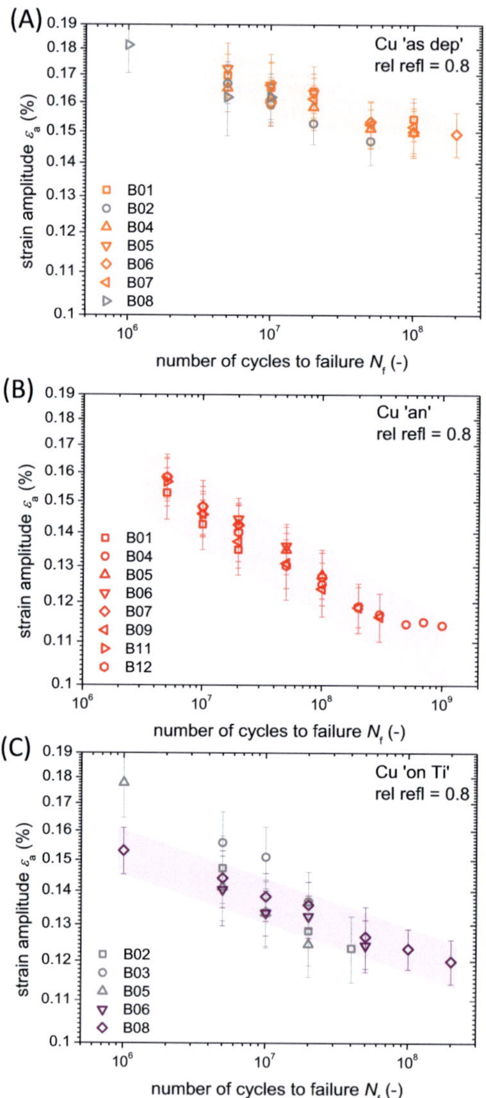

Fig. 5.20: Lifetime diagrams for all successfully tested (A) 'as deposited', (B) 'annealed', and (C) 'on Ti' Cu thin films with a failure criterion of a relative reflectivity of 0.8. The grey symbols indicate difficulties occurred during the experiment.

A summary of all trusted Cu thin film lifetime data and the fitted Basquin Law (Eq. 2.2) is shown in Fig. 5.21. The finer grained 'as deposited' Cu thin films show the highest lifetimes and a fatigue sensitivity of $b = -0.04$. The coarser grained 'annealed' Cu thin films display lower lifetimes and a higher sensitivity with $b = -0.07$ than the 'as deposited' ones. For very high cycle numbers up to 10^9 a plateau can be observed. The fine-grained Cu thin films on the Ti seed layer have further reduced lifetimes but with approximately the same sensitivity $b = -0.05$ as the 'as deposited' thin films. The fatigue data is summarized in Table 5.4.

Lifetime diagrams of selected cantilevers with different failure criteria are plotted in Fig. 5.22 for all three types of Cu thin films. The failure criteria range from a relative reflectivity of 0.9 as conservative criterion down to 0.5 as non-conservative criterion. In general, the trend of the lifetime curves remains but is shifted towards lower lifetimes for the conservative criterion.

Table 5.4: Fatigue data of Cu thin films.

sample	fatigue sensitivity factor b	fatigue strength σ_f (MPa)
Cu 'as dep'	-0.04	358
Cu 'an'	-0.07	514
Cu 'on Ti'	-0.05	355

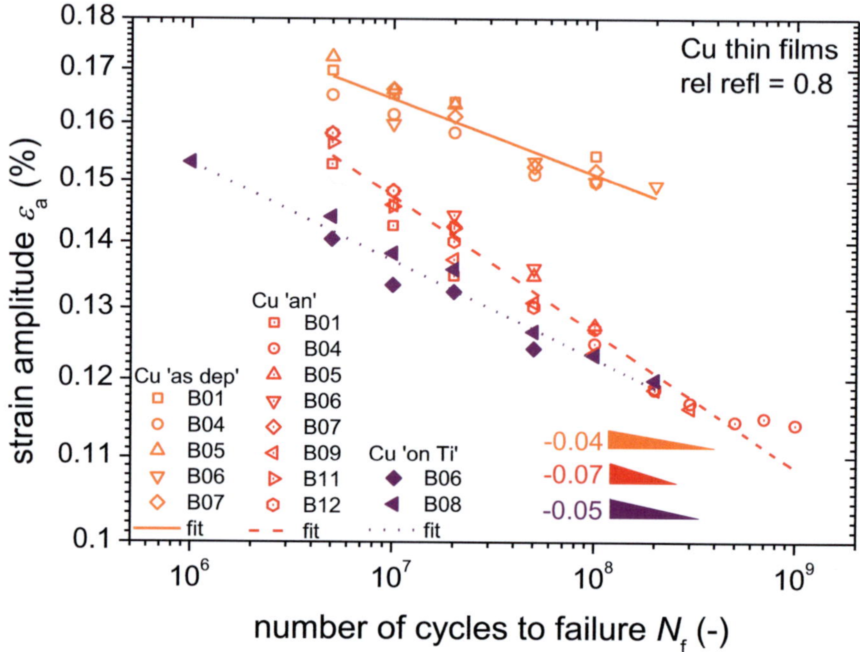

Fig. 5.21: Lifetime diagram for 'as deposited', 'annealed' and 'on Ti' Cu thin films.

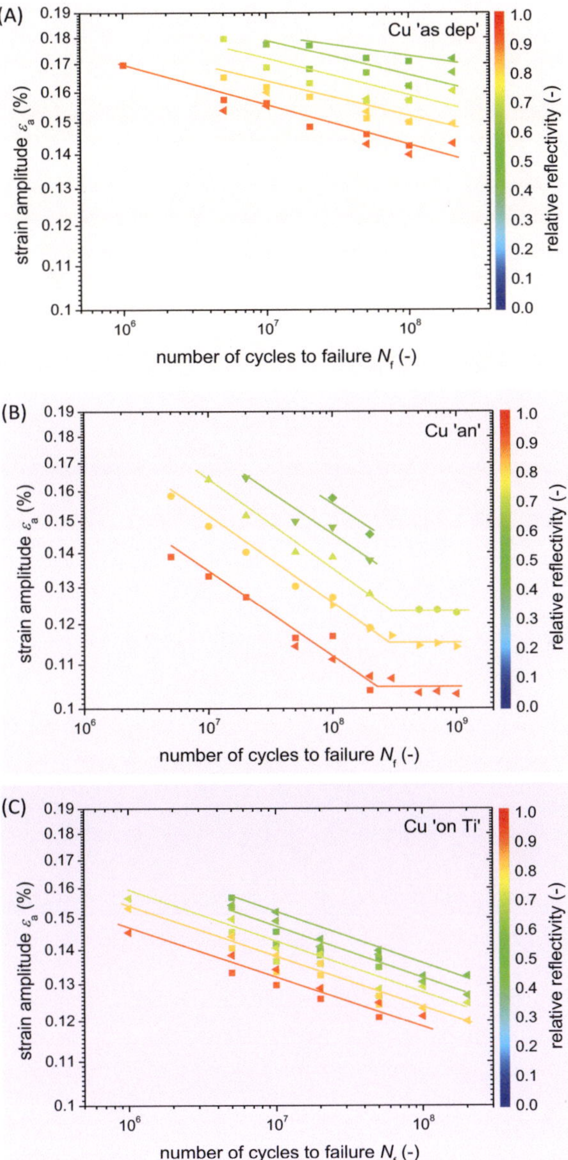

Fig. 5.22: Lifetime diagrams of (A) 'as deposited', (B) 'annealed', (C) Cu 'on Ti' Cu thin films for several failure criteria.

6. Discussion

In this chapter the experimental results will be discussed. First, the damage mechanisms in Al and Cu thin films are proposed, followed by a more detailed and separate explanation of damage formation in either Al or Cu thin films. Then, the lifetime data of the Al and Cu thin films will be compared to each other and to literature data. Table 6.1 summarizes the literature used to compare damage formation (DF) or lifetime data (LD). In the end a phenomenological lifetime model is presented.

Table 6.1: Overview of literature data of other thin film samples.

film	sub-strate	*t* (nm)	*d* (nm)	R	*f* (Hz)	re-gime	criterion		ref
Cu	free	1100	980	0.1	0.07	LCF	amplitude	LD	[74]
Cu	poly	3000	800	0.1	<1	LCF	strain range	DF	[113]
Cu	poly	400-3100	280-810	-1	0.1-0.3	LCF	energy loss	LD	[114]
Cu	poly	50-3000	110-1630	-1	100	HCF	extrusions	DF	[80]
Cu	poly	50-128	20-30	0.05	10	LCF	energy loss	DF	[89]
Cu	poly	100-3750	50-600	0.05	5	LCF	resistance	LD	[81]
Cu	Si	100+300	300+1500	therm	100	HCF	electrical	LD DF	[115]
Cu	micro	3000000	200	-1	1	LCF	fracture	DF	[116]
Cu/ Nb	free	40000	40 layers	-1	800	HCF	fracture	LD	[83]
Al	free	2250	-	0.1	0.02	LCF	fracture	LD	[7]
Al	free	1100	-	0.13	100	HCF	fracture	LD	[84]
Al	poly	80-800	50-300	0.05	5	LCF	resistance	LD	[81]
Al	poly	400	-	biaxial	0.2	LCF	optical	LD	[87]
Al	Si	420	300	-	900M	UHCF	frequency	LD DF	[79]
Al	micro	5000000	350-850	-1	130	VHCF	fracture	LD DF	[117]
Ag	Si	600	1000	-	45	HCF	stiffness	DF	[93]
Au	poly	400	50	biaxial	0.2	LCF	optical	LD	[87]

6.1. Damage Mechanisms in Al and Cu Thin Films

This chapter introduces the proposed deformation mechanisms involved in the fatigue induced damage process in Al and Cu thin films. The main differences between Al and Cu are the damage feature types (hillocks, extrusion islands, pores and short cracks) and their distribution throughout the sample surface. Furthermore, there is an effect of initial grain size on damage formation independent of the material. The Ti seed layer has a pronounced influence on damage in the Cu thin films. Therefore, the discussion will be opened by a summary of microstructure and damage features of the tested Al and Cu thin films and their different conditions. A summary of microstructural features obtained in the thin films can be found in Table 6.2 and the proposed damage mechanisms are displayed in Fig. 6.1.

The microstructural damage morphology of Al thin films can be summarized to contain extrusion islands, hillocks, and short cracks at the thin film surface (Fig. 5.4) and pore formation within the cross section (Fig. 5.2 (B), Fig. 5.3 (B)). In both Al thin film conditions it was found that the film thickness loses its uniformity and varies locally in thickness. This supports the influence of diffusion processes, due to the high homologous temperature of Al. The hillock formation can be associated with diffusion processes at grain boundaries [59] and seems to be a reasonable process to release locally accumulated compressive stresses during cycling. The formation of extrusions and pores can be related to the annihilation of dislocations and diffusion of vacancies, described by Schwaiger et al. [90]. Additionally, the surface passivation layer might have an effect on damage formation. The short cracks visible in the SEM micrographs (Fig. 5.2 (A), Fig. 5.3 (A)) are believed to have been caused by tensile stresses at grain boundaries, leading to a stress relief.

Cu thin films only show extrusion islands and short cracks at the surface (Fig. 5.15) as well as pore formation within the cross section (Fig. 5.12 (B), Fig. 5.13 (B), Fig. 5.14 (B)). Compared to the Al, the diffusivity of Cu is low at room temperature, due to its higher melting point. Therefore, hillock formation is kinetically not favored. Extrusion and pore formation can be attributed to the dislocation processes mentioned earlier. The short cracks may be related to a stress relief in between damaged areas.

In Al the fatigue damage is distributed homogeneously throughout the sample surface while in the case of Cu there are damaged and undamaged regions (compare Fig. 5.4 and Fig. 5.15). This is assumed to be due to the elastic isotropy of Al and elastic anisotropy of Cu.

For both thin film materials, the 'as deposited' thin films with a relatively fine-grained initial microstructure show grain growth. As mentioned before, the homologous temperature for Al is higher than for Cu. Therefore, diffusion processes may take place in the Al thin films but much slower in the Cu thin films. In the case of Cu the grain growth may be induced by a coupling process proposed by Cahn et al. [54].

A more detailed perspective of each thin film condition is discussed separately in the following chapters.

Table 6.2: Summary of microstructural features of Al and Cu thin films after around 10^8 cycles and strain amplitudes of $\varepsilon_a^{max} = 0.17 - 0.20$ %.

	Al 'as dep'	Al 'an'	Cu 'as dep'	Cu 'an'	Cu 'on Ti'
initial *d* (nm)	800	1200	200	1200	400
grain shape	globular	columnar	globular	columnar	globular
in-plane texture	no	no	no	no	yes
heat treatment	-	1 h @ 450 °C in UHV	-	1 h @ 450 °C in UHV	-
surface	passivation	passivation	free	oxidized	free
interface	Al/Si	Al/Si	Cu/Si	Cu/Si	Cu/Ti
***d* (nm) after fatigue**	1200	1200	1200	1200	1200
median fatigue featuresize (nm)	2400	2400	3000	3000	5000
extrusion morphology	coarse parallel 45 ° oriented	fine parallel random	coarse parallel 45 ° oriented	fine parallel random	super fine random irregular shape
pore morphology	cross section interface	cross section	cross section interface	cross section interface	near surface
crack morphology	at GBs	at GBs	-	at GBs	at extrusion lamellas
hillocks	yes	yes	no	no	no

mechanism	grain growth	hillock formation	extrusion formation	pore formation	short cracks
Al 'as dep'					
Al 'an'					
Cu 'as dep'					
Cu 'an'					
Cu 'on Ti'					

Fig. 6.1: Summary of proposed damage mechanisms in Al and Cu thin film sample.

6.2. Damage Formation in Al

The results of cycled Al thin films presented in chapter 5.2 will be discussed with respect to the initial microstructure, the mechanisms of fatigue damage formation, and the quantitative fatigue damage analysis.

6.2.1. Initial Microstructure of Al Thin Films

The initial microstructure of the thin film has a large effect on the damage formation during cycling. The initial grain size of the 'as deposited' Al thin films of 800 nm is relatively small (see Fig. 5.1 (A)). As the sputter process is very sensitive to the deposition parameters, a comparison between different sputter setups is difficult. Nevertheless, the small grain size probably results from the fact that the substrate was not heated during deposition. The grooves and single hillocks of the 'as deposited' Al thin films may be due to the elevated base pressure. Additionally, the incorporation of impurities might have led to the fine-grained structure.

The 'annealed' Al thin films have a grain size of around 1200 nm, which is caused by grain growth during annealing (see Fig. 5.1 (B)). The heat treatment was not conducted directly after the sputter process but after the vacuum chain was broken and a natural passivation layer on the Al surface was formed. This might be the reason why the initial surface morphology can be observed on the 'annealed' thin films. During annealing, the thin film formed hillocks and pores. The hillock formation observed after annealing in thin films is well known from literature and is attributed to the formation of compressive stresses during thermal treatment [57]. The pore formation at the surface can be attributed to the material transport during annealing to form the hillocks.

6.2.2. Damage Formation Mechanisms in Al Thin Films

The damage formation after a specific cycle number at different strain amplitudes for both Al thin film conditions is compared in Fig. 5.4. It can be noted that both Al thin film conditions show hillocks, extrusion islands and short cracks at the surface and pores either at the interface with the substrate or within the cross section (see Fig. 5.2 (B), Fig. 5.3 (B)). Differences in the damage formation can be attributed to different damage mechanisms due to the initial microstructure.

Distribution of Damage

As stated before, the damage formation for the 'as deposited' and 'annealed' Al thin films is phenomenologically very similar. In both cases the damage is homogeneously distributed throughout the surface (Fig. 5.4). Therefore, it is proposed that the grain orientation does not have a strong effect on the damage formation, which can be explained by the elastic isotropy of Al crystals. The elastic isotropy leads to the same stress amplitudes in all the grains independent of their orientation, which might result in a more homogenous damage formation. This is in contrast to the orientation dependence of the damage formation in Ag thin films observed by Schwaiger et al. [93]. They found that damage occurs preferentially in (100)-oriented grains, and attributed that to a lower flow stress compared to (111)-oriented grains.

Grain growth

In the 'as deposited' Al thin films, cyclic induced grain growth takes place within the first 10^6 cycles prior to damage formation. Mechanically induced grain growth in nanocrystalline Al thin films was observed e.g. by Gianola [118]. The authors attributed the grain growth to stress coupled grain boundary migration. As a grain boundary consists of dislocations, it can move due to an applied stress on the boundary. Another possible mechanism of grain growth could be diffusion based processes at the grain boundary. As the homologous temperature of Al at room temperature is about a third of the melting temperature, diffusion processes could occur and help in the migration of grain boundaries. The small grain size leads to an enhanced diffusivity and therefore diffusion processes are expected to strongly influence fatigue induced damage formation.

Hillock Formation

Both Al thin films display hillocks after cycling, whereas in the 'annealed' thin films the hillocks partially result from the heat treatment. As the thin films are subjected to fully reversed loading, there could be cyclic induced compressive stresses due to irreversible dislocation glide. These induced compressive stresses might cause the hillock formation. Looking at the cross sections (Fig. 5.2 (B), Fig. 5.3 (B)), the hillocks seem to consist of one grain that is rotated out of the surface. This is in contrast to the findings of Kim et al. [60]. They observed a different microstructure of the hillock that forms underneath the thin film and pushes the film up (see Fig. 2.2 (B)). Alternatively, the comparison with electromigration induced hillock formation in Al thin films could be drawn. Two possible processes were proposed by Nucci et al. [61]. Either the grain rotates first and is then pushed out by grain boundary sliding, or it is an alternating process where both processes take place simultaneously, which would result in a round shape of hillocks (see Fig. 2.2 (C), (D)). Although there is no clear prove it is believed that the latter mechanism takes place in the fatigue induced hillock formation of the Al thin films investigated here. Whereas in the 'as deposited' Al thin films diffusion might have an additional influence, in the 'annealed' thin films it seems to be the passivation layer.

Extrusion Formation

Subsequent to the hillock formation, extrusion islands are formed. The shape and size gives rise to the assumption that they correspond roughly to the grain structure. The extrusion formation seems to be caused by dislocation processes, as proposed by Zhang et al. [91] for Cu thin films. The extrusion lamellas in the 'as deposited' Al thin film are coarser but oriented in 45 ° to bending direction (Fig. 5.2 (A)), whereas the 'annealed' Al thin films display finer extrusion lamellas with random orientations (Fig. 5.3 (A)). This difference could be due to the initial grain size.

In the 'as deposited' case, the first stage is believed to be grain growth, where the grains with a preferred crystal orientation would be favored. Since the 'as deposited' films have a finer grain structure there is a higher chance that the right orientation is available in a large enough grain. Therefore, more grains could be damaged in single slip orientation with a high Schmid factor leading to a stronger preferred orientation of the extrusion lamellas.

The 'annealed' thin films have a large enough volume to allow dislocation processes from the beginning of the cycling. Due to the annealing and following cooling step, the dislocation density is not reduced in the 'annealed' thin films. Dislocations could form easily on their preferred glide planes. The random orientations of the extrusion lamellas might be due to different in-plane orientations of the grains. Furthermore, the formed passivation layer might have an additional effect on the lamella spacing.

Pore Formation

In the 'as deposited' Al thin films the pores are found at the interface to the substrate or within the cross section (Fig. 5.2 (B)). They could be formed similarly to the model proposed for Cu thin films by Schwaiger et al. [90], where dislocation annihilation as well as vacancy diffusion are proposed to result in pores. Although, volume diffusion is rather slow, diffusion of vacancies to sinks at interfaces or grain boundaries may still have an influence on the pore formation.

The 'annealed' thin films display pores within the cross section (Fig. 5.3 (B)), but are not found along potential slip lines as in the Cu thin films. Besides dislocation annihilation and vacancy diffusion, the passivation layer seems to have an influence. Furthermore, it could be argued that due to hillock formation and the original surface is rotated into the thin film. Therefore, the pores might be transported into the cross section due to grain rotation.

Crack Formation

Short cracks are observed in both Al thin film conditions and occur in between damaged areas (Fig. 5.2 (A), Fig. 5.3 (A)). Localized tensile stresses due to different damage formation and orientation might have developed during cycling and are relieved by local cracking. It has to be noted that crack nucleation in a thin film on a substrate does not necessarily lead to a fatal failure. Therefore the short cracks observed here are localized damage features and do not propagate as in bulk material under cyclic loading.

Comparison to Other Damage Structures

In this section the beforehand described damage formation of the Al thin films tested here are compared to fatigue of Al thin films from literature. Grain growth, short cracks and something similar to hillock formation was also reported by Eberl et al. [79]. They tested Al thin films with surface acoustic waves (SAW) up to 10^{12} cycles at a strain amplitude of 0.03 %. They found voids at grain boundaries and triple points, which is somehow in contrast to the pore formation in the here studied Al thin films. Additionally, the extrusion formation differs, as they observed single high aspect ratio extrusions at the surface protruding from grain boundaries instead of extrusion islands with parallel lamellas. The differences of the two damage formation structures may be related to the large differences in testing conditions. The Al thin films tested in this thesis were fatigued at 500 Hz, while the SAW test devices were working at 1 GHz. Therefore, the frequency and applied strain rates are orders of magnitude apart. Furthermore, the geometric differences could have an influence on damage formation. The electrode lines in the SAW devices have a width of 1 µm and the additional free surfaces at the side walls offer opportunities for the formation of high aspect ratio extrusions.

For bulk materials, Höppel et al. [117] observed single extrusions on the surface of ultra fine-grained Al micro samples and a reduced roughening compared to coarse-grained counterparts. By milling FIB cross sections at extrusion sites they observed subsurface void formations similar to what was found in the thin film samples. They attributed these pores to the condensation of vacancies introduced by the sample preparation process. The volume of these micro samples is much larger than in the thin films observed here, but the grain sizes are comparable. While the pores in the micro samples seem to be similar to the ones observed in thin films after fatigue, but the fabrication processes are rather different. Therefore, the interpretation of their cause might be rather different and in thin films purely related to the fatigue damage.

6.2.3. Quantitative Damage Analysis of Al Thin Films

The quantitative analysis of the Al thin films, presented in chapter 5.2.2, approves the optical characterization of the thin film surfaces by the reflectivity of the laser. The fraction of undamaged area corresponds to the trend of the reflectivity within the accuracy of the measurement (see Fig. 5.5). The roughness shows the opposite trend, as a low fraction of undamaged surface corresponds to a high roughness. Thus, three different measurements (reflectivity, fraction of undamaged area, and roughness) show the same trend for the surface evolution.

This confirms the applicability of the reflectivity measurement, which is experimentally easy to access.

The line and area analysis reveal that the number of damage features increases with increasing strain amplitude until saturation is reached. The fatigue feature size also increases with the strain amplitude (see Fig. 5.7). For both Al thin film conditions, the maximum fatigue feature size is around 2.4 μm. The 'annealed' thin films reach this saturation size at lower strain amplitudes. This is probably due to the larger initial grain size of the 'annealed' Al thin films compared to the 'as deposited' films. For the 'as deposited' thin films grain growth takes place prior to damage formation, which might cause the shift towards higher amplitudes. The maximum fatigue feature size is determined by geometrical aspects, as it can be imagined that the damaged areas may not impinge on each other. The number of cycles to reach the fatigue feature size does not have a pronounced impact. This implies that the nucleation of damaged features is the limiting factor, while the formation of the damage in the grains is much faster, and cannot be observed by the method. For a better insight what happens in detail in terms of damage nucleation and formation, further investigations have to be conducted.

6.3. Damage Formation in Cu

In this chapter the damage formation and mechanisms for the 'as deposited', 'annealed' and 'on Ti' Cu thin films are discussed based on the results presented in chapter 5.3 with respect to the initial microstructure, the mechanisms of damage formation, and the qualitative analysis of the damage formation.

6.3.1. Initial Microstructure of Cu Thin Films

The initial microstructure of the Cu thin films has a strong impact on lifetime and damage formation during fatigue. All Cu thin films were sputter deposited without heating the substrate. The 'as deposited' Cu thin films have a fine-grained microstructure and show grooves at the surface (Fig. 5.11 (A)). Due to the rather high base pressure a higher excess volume can be expected and the low substrate temperature leads to a reduced mobility of adatoms which cannot compensate the shadowing effects created through the non-rotating substrate [119]. The FIB micrograph displays different grey values of the grains by channeling contrast and indicates no specific in-plane orientation. This argument is supported by the XRD diffraction pattern (Fig. 5.16 (A)), which gives rise to a fine and random microstructure as the pattern shows continuous Debye-Scherrer rings with no in-plane orientation. Furthermore, to use a direct method to obtain the local microstructure, TEM investigations were carried out. The TEM micrograph in Fig. 5.11 (A) shows grains on the order of 200 nm with a certain amount of growth twins.

The 'annealed' Cu thin films might have formed a surface oxide as the vacuum chain was broken between deposition and heat treatment. This explains why the 'annealed' Cu surface shows grooves from the initial deposition (Fig. 5.11 (B)). Furthermore, the 'annealed' Cu thin films show pores at the surface and single hillocks. The FIB channeling micrograph reveals much larger grain sizes on the order of 1.2 μm with no in-plane orientation. The grains contain annealing twins which is in good agreement with the TEM characterization. The XRD diffraction pattern (Fig. 5.16 (B)) shows incomplete Debye-Scherrer rings with individual high intensity spots. This suggests an enlarged grain size due to annealing.

The Cu thin films deposited 'on Ti' seed layer display coarser surface features (Fig. 5.11 (C)). However, the channeling contrast in the FIB micrograph reveals the fine-grained structure with a high amount of light grey grains. This indicates an in-plane texture which is confirmed by the XRD diffraction pattern in Fig. 5.16 (C), as it shows partial Debye-Scherrer rings. The TEM micrographs show a fine-grained microstructure as well.

6.3.2. Damage Formation Mechanisms in Cu Thin Films

In the following chapter the damage formation and the underlying mechanisms of Cu thin films are discussed. The damage features in the Cu thin films consist mainly of extrusions. They are sometimes accompanied by short cracks at grain boundaries and pore formation within the thin film. To understand the influence of the grain size and the interface to the substrate, the damage evolution of the three different types of Cu thin films will be compared in the following paragraph.

Distribution of Damage

The damage at the surface of the Cu thin films is inhomogeneously distributed, with damaged and undamaged areas (Fig. 5.15). This could be related to the high elastic anisotropy of Cu. Due to the elastic anisotropy, each grain in the thin film undergoes a different stress depending on its orientation relative to the applied strain amplitude, which is provided by the substrate. Therefore, the grains with high stress amplitudes will preferentially form extrusions.

Grain Growth

It is argued that grain growth takes place during fatigue in the Cu 'as deposited' and Cu 'on Ti' thin films. This is supported by TEM micrographs (Fig. 5.12 (C), Fig. 5.14 (C)), which show larger grains than before fatigue (Fig. 5.11 (A)+(C)). Additionally, the XRD investigations support this observation, because it was found that the continuous Debye-Scherrer rings observed in the unfatigued regime begin to dissolve into single spots in the highly fatigued areas (Fig. 5.16 (A)+(C)). The dissolution of a continuous ring to individual high intensity spots can be attributed to grain growth. As the homologous temperature of Cu at room temperature is less than a third of the melting temperature, diffusion processes are unlikely to happen although it has to be noted that the concentration of vacancies might be largely enhanced due to the severe damage formation. However, there is something like a stress or strain assisted grain growth occurring in the fine-grained Cu thin films which takes place within in the first 10^6 cycles. The process of grain growth in the Cu thin films might be similar to a coupling process of grain boundary motion proposed by Cahn et al. [54].

Extrusion Formation

The extrusion formation in Cu thin films can be attributed to the irreversible glide of dislocations on favored glide planes. The presence of dislocations is confirmed by the TEM micrographs of 'annealed' and 'on Ti' Cu thin films (Fig. 5.13 (C), Fig. 5.14 (C)). No distinct dislocation structures with walls and cells as in bulk material could be identified. Instead, loosely arranged dislocation structures like tangled dislocations were found, as also observed by Zhang et al. [91] for fatigued Cu thin films in the LCF regime. The absence of high ordered dislocation structures can be attributed to the dimensional constraints on dislocation nucleation and motion in thin films.

The median fatigue feature size of the 'as deposited' and 'annealed' Cu thin films are around 3 µm and for Cu 'on Ti' around 5 µm. Contrary, the grain sizes after cycling range around 1200 nm for all different conditions of Cu thin films (Table 6.2). The extrusion formation seems to span over several grains, which is more pronounced in the Cu 'on Ti' thin films. This in turn, can be attributed to the in-plane texture of Cu 'on Ti' thin films observed by XRD (Fig. 5.16 (C)). If adjacent grains have similar orientations it might be easier to form extrusions spanning over several grains and result in a larger median fatigue feature size compared to the other Cu thin films. Furthermore, it was observed that the in-plane texture of Cu 'on Ti' thin films changes, due to the shift in position of the partial Debye-Scherrer rings (Fig. 5.16 (C)), which could be a sign for grain rotation [120]. But this process might not be reasonable, because of the Ti adhesion layer. It might be argued that the change in in-plane texture corresponds to an enhanced growth of a certain type of grain orientation, where extrusions can form. Grains with slip systems with high Schmid factors may be preferred. It might also be possible that the extruded surface and a small change in orientation due to the extrusion lamellas itself, causes a shift of the partial Debye-Scherrer rings.

The extrusion lamellas in the 'as deposited' thin films often show a 45 ° orientation relative to the bending direction (Fig. 5.12 (A)). Whereas, the extrusion lamellas of the 'annealed' Cu thin films seem to be randomly oriented, but are parallel within an extrusion island (Fig. 5.13 (A)). In contrast, the extrusion lamellas in the Cu 'on Ti' thin films are very fine, randomly orientation and not parallel within an extrusion island (Fig. 5.14 (A)). The different forms of extrusion lamellas might result from combined effects of grain growth, grain orientation, and the grain shape. Generally, extrusions are expected to form first on glide planes with a high Schmid factor. The orientation of these glide planes in turn, depends on the orientation of the grain. In the case of 'as deposited' Cu thin films the cyclic induced grain growth seems to favor certain orientations and a change from globular to columnar shaped grains might lead to the oriented extrusion lamellas. As the microstructure of 'annealed' Cu thin films is stable during cycling, the extrusion lamellas form accordingly to the initially given microstructure. For the Cu 'on Ti' thin films the extrusion lamella formation is even more complex as the Ti layer seem to have a pronounced effect, especially on the size of the lamellas. The strong interface between Cu and Ti does not seem to allow dislocations piercing through which would lead to a backstress on the dislocation sources, similar as described in [55, 56]. Such a backstress might be able to constrain the dislocation sources. A continued damage formation would then only be enabled when other dislocation sources were activated. The result would be that multiple sources could only emit a limited amount of dislocations leading to a more continuous and fine extrusion lamella formation. It is assumed due to the appearance of the fine lamellas that a lot of single slip on parallel glide planes is happening rather than multi-slip on glide systems with similar Schmid factors.

Pore Formation

Pore formation is found in all three types of Cu thin films but the pores are different in their appearance. In the 'as deposited' and 'annealed' Cu thin films, pores formed at the interface with the substrate as well as within the cross section (Fig. 5.12 (B), Fig. 5.13 (B)). The formation is more pronounced in the 'as deposited' thin films. The pores decorate lines in the cross sections which might correspond to glide planes in the grain. The pore formation seems to be similar to the process proposed by Schwaiger in [75], where the annihilation of dislocations of opposite signs leads to vacancy formation, which accumulate to pores during further loading. The position of the pores at the interface furthermore corresponds to the position of the extrusion lamellas. As the film thickness stays overall the same but is only kind of sheared off, diffusion processes with long range material transport are not likely to happen.

The pore structure in the Cu 'on Ti' looks very different (Fig. 5.14 (B)). There are no pores at the interface, but a high pore density is found near the surface of the Cu thin film. The pores also align along certain lines. This phenomenon can be explained by the effect of the Ti seed layer. On the one hand, Ti increases the adhesion of the Cu thin films, which was also shown in the case of a Ta layer beneath a Cu thin film [55]. The effect of such a layer was also simulated by Nicola et al. [56]. They found that dislocations pile up at the interface and create a backstress. Therefore, it was argued above that the dislocations which pile up at the interface can inhibit further dislocation nucleation at a given source. Such a pile-up would lead to a high dislocation density at the interface and would promote dislocation annihilation. According to Essmann et al. [69] and Schwaiger et al. [75, 90] this would lead to pore formation. Since this cannot be observed in the Cu 'on Ti' thin films, one of the necessary processes seems to be inhibited. As the Ti is believed to be a strong interlayer, the dislocation pile-up might be stable enough to pin dislocations in a way that no annihilation occurs. Furthermore, it seems that the Ti reduces diffusion processes which would also be necessary for pore formation. Overall, these assumptions would create a stable backstress which pushes all the dislocation processes towards the Cu surface, where enhanced annihilation of dislocations can take place and leads to a high pore density.

Crack Formation

The fatigued Cu thin films display short cracks between adjacent extrusion islands. These cracks appear preferably in 'annealed' and 'on Ti' Cu thin films. Such features might be caused by local tensile stress release due to different orientations of damaged areas.

Comparison to Other Damage Structures

The beforehand described damage features of the here investigated Cu thin films can be compared to fatigued Cu thin films reported in the literature. The microstructural damage investigations on Cu thin films on polymer substrates in the LCF regime can be found for example in [80, 81, 88, 90, 91, 113]. The main focus of these studies was the observation of size effects in the fatigue damage of Cu thin films.

The fine-grained 'as deposited' Cu thin films show grain growth and extrusion formation (Fig. 5.12). This is in contrast to what was found for cycled Cu thin films with small grains in the LCF regime, reported by Zhang et al [91]. They did not observe grain growth, but mainly found roughening of the Cu thin film and an enhanced grain boundary cracking. Extrusions were only observed within some larger grains. These major differences could be due to the testing conditions. In the LCF regime higher amplitudes were used leading to less cycles to failure. Therefore, it can be assumed that the first 10^6 cycles at low elastic strain amplitudes are dominated by grain growth and if the grains are large enough they begin to form extrusions.

The 'annealed' Cu thin films display extrusions, pores and tangled dislocation structures (Fig. 5.13). This is similar to the investigations of 1 µm Cu thin films with large grains on polymer substrate in the LCF regime conducted by Zhang et al. [91]. They observed extrusion islands at the surface and pore formation within the cross section of the film. Their TEM investigations revealed tangled dislocation structures to be the cause of damage formation. In thermally cycled Cu lines from Park et al. [94] or Mönig [115] in the HCF regime, they observed extensive grain growth with extrusion formation. Similar to what is found in the study presented here, the thermally cycled Cu thin films have damaged and undamaged regions and twin boundaries additionally can hinder or promote damage formation depending on the orientation.

6.3.3. Quantitative Damage Analysis of Cu Thin Films

The comparison of the quantitative analysis between reflectivity and the undamaged area fraction underlines the applicability of the reflectivity as a damage criterion (Fig. 5.17). The line analysis of the damage areas allows measurement of the number and the size of the fatigue damage (Fig. 5.19). In all cases, the number of extrusion areas saturates with increasing strain amplitude. The 'annealed' Cu thin films show the highest density of damaged islands while the 'as deposited' shows a lower density and Cu 'on Ti' thin films has the lowest.

The median sizes of the damage features are the same for the 'as deposited' and 'annealed' Cu thin films for the same strain amplitude. This correlates with the lower lifetime of 'annealed' Cu thin films observed in the lifetime diagram (Fig. 5.21), as the 'annealed' display a higher damage density. While the grains in the 'as deposited' thin films might be too small to allow an extensive extrusion formation right from the beginning, this is not the case for the 'annealed' ones. Instead, the 'as deposited' thin films first have to undergo grain growth. The Cu 'on Ti' shows more damage islands with a median damage size of over 5 µm. This could be attributed to the in-plane orientation of the Cu 'on Ti' and a strongly promoted damage formation ranging over several grains. Therefore, the extrusion formation might be easier as the grains have similar orientations.

The fact that the cycle number does not seem to have an effect on the damage feature size might be similar what was proposed for the Al thin films. The nucleation of the damage features seem to be the limiting factor depending on the amplitude, whereas the damage formation itself might be occurring to fast to be captured by the cycling steps.

6.4. Lifetime Data in Comparison

In this chapter the lifetime data obtained by using the reflectivity criterion are compared for the Al and Cu thin films tested, as well as to literature data.

6.4.1. Comparison of Lifetime of Al and Cu Thin Films

The lifetime diagrams obtained for the different Al and Cu thin films are compared to each other in Fig. 6.2. The failure criterion was a relative reflectivity of 0.8. In general, the Cu thin films display longer lifetimes than the Al, which can be explained by the higher flow stresses of the Cu thin films in the case of monotonic loading.

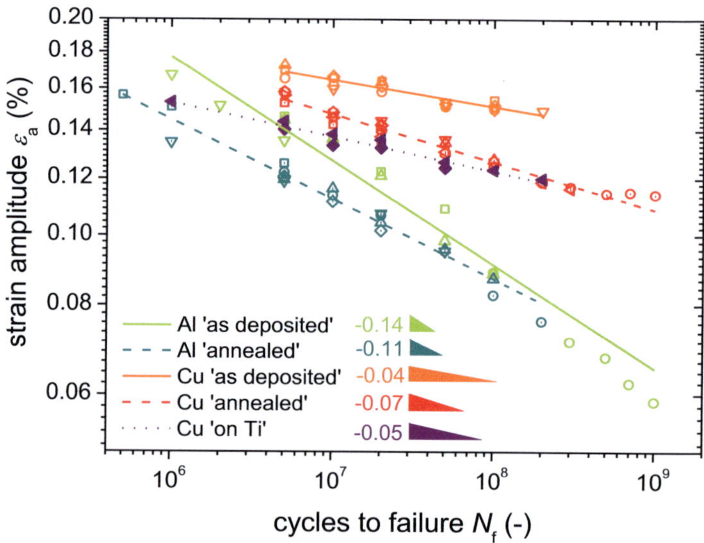

Fig. 6.2: S-N curves of Al and Cu thin films for a failure criterion of relative reflectivity is 0.8.

Furthermore, the Al thin films have a higher sensitivity to fatigue than the Cu thin films. This means that the dependence of the thin film lifetime on the strain amplitude is stronger for the Al than for Cu. This is somewhat unexpected, as the natural passivation layer of Al thin films could be seen as a constraint to dislocation processes which would slow down the fatigue damage formation. But, the high homologous temperature of Al promotes diffusion processes, which seem to account for enhanced damage formation. The additionally happening diffusion process would increase the fatigue damage feature formation, which would lead to the higher sensitivity of Al.

For both materials, the initially fine-grained 'as deposited' thin films show longer lifetimes than the 'annealed' thin films. This is just what is expected from grain hardening, e.g. Hall-Petch [34, 35] and the model of Thompson [36], under the monotonic loading conditions, and seem to be transferrable into the cyclic loading case, at least for the HCF regime.

For the Al thin films the fatigue sensitivity factor is higher for the 'as deposited' films than for the 'annealed' films. With increasing cycle numbers, the grain growth during cycling leads to a similar grain size in 'as deposited' and 'annealed' Al thin films. Therefore, the lifetime becomes similar for very high cycle numbers. The fatigue at higher strain amplitudes seems to occur later for 'as deposited' films due to the time dependent grain growth mechanisms. The Al thin films do not show an endurance limit.

For the Cu films, the fatigue sensitivity of 'as deposited' thin films and Cu 'on Ti' is similar, and lower than for the 'annealed' Cu thin film. For very high cycle numbers of 10^9 the 'annealed' Cu thin films seem to display a limit. This is not expected as, typically, pure fcc materials do not tend to display an endurance limit and the existence of a limit is still controversial [67]. It might be argued that the limit observed for the Cu thin films results from a locally decreased strain amplitude due to the beforehand occurred damage. This should be unlikely as the substrate carries the load and induces the strain into the films locally. Unfortunately, the influence of adjacently damaged areas on the afterwards occurring damage is not clear. There might be a threshold value or incubation time for damage nucleation in the case of Cu thin films below a certain strain amplitude. Further investigations with even longer cycle numbers are necessary to understand the fatigue behavior in this case. The Cu 'on Ti' thin films show the lowest lifetimes, which can be understood by extensive extrusion formation that happens at the surface. The Ti seed layer therefore does not have a strengthening effect but affects the dislocation mechanisms and pushes them towards the surface, where damage is caused as discussed previously. It has to be kept in mind that the failure criterion used is surface sensitive. Since the damage in Cu 'on Ti' is concentrated at the surface, this might give the impression of a reduced lifetime compared to other failure criteria, e.g. resistance measurements.

6.4.2. Comparison with Literature

In this chapter the lifetime data obtained here is compared to selected fatigue data of Al and Cu samples summarized in Table 6.1 and presented in Fig. 6.3. Unfortunately, there is no direct comparison to the data observed here. It has to be noted that the thin film data on Si substrate in the HCF regime is compared to LCF experiments, thin films on polymer substrate, freestanding thin films, and micro samples with a comparable grain size. Also, the test frequencies and the R ratio differ as well as the damage criterion and the test temperature, e.g. thermally cycled thin films. Although all these parameters will have an impact on lifetime the available literature data is too scarce to leave them out.

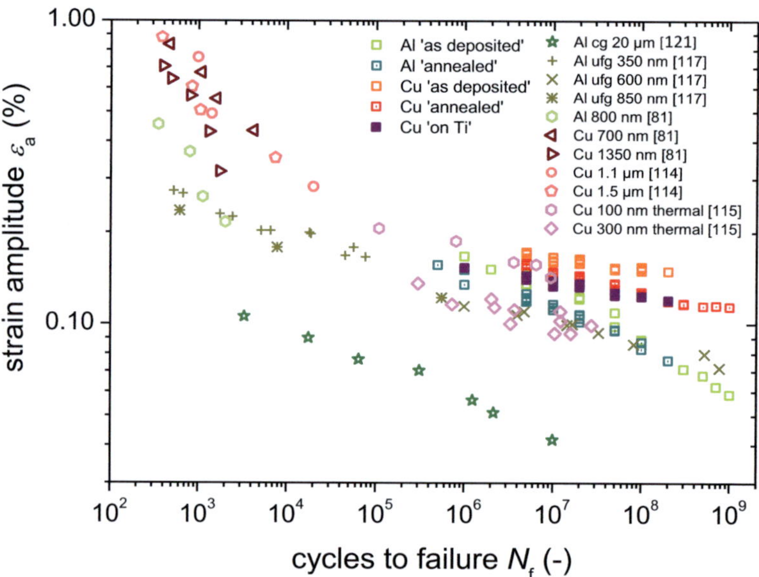

Fig. 6.3: S-N curves for Al and Cu thin films from this work in comparison with literature data. It has to be noted that different measurement methods are compared. See Table 6.1 for details.

For better comparison, the lifetime data is plotted as total strain amplitude, which in the case of samples tested in the LCF regime is the plastic strain amplitude, and in the HCF it is the elastic strain amplitude. All differences aside, some trends can be identified. At least two regimes can be distinguished for the small scale data. There is the LCF regime up to 10^4 cycles and the HCF regime up to 10^8 cycles. The behavior in the VHCF regime for more than 10^8 cycles differs for the Al and Cu samples.

For Al samples a sequence can be plotted bridging sample sizes from macro to micro. Bulk macro samples with large grains of 20 µm [121] have the lowest lifetimes. The lifetime is increased if the grain size is reduced below 1 µm [117]. In the LCF regime there is a slight increase in lifetime if Al thin films on polymer substrate [81] are compared to ufg materials. This is true in the HCF regime, where Al thin films show a higher lifetime.

For Cu samples an LCF [81, 114] and an HCF regime can be distinguished. In the HCF regime only thermally cycled Cu thin films are available [115]. Surprisingly, the thinner 100 nm and 300 nm Cu films display a comparable or even lower lifetime than the 1 µm thin Cu films from this study. This can be attributed to the higher temperature during thermal cycling. With the higher temperatures, diffusion processes are activated in the Cu thin films.

6.5. Phenomenological Lifetime Model

In this paragraph a fatigue feature growth model is used to describe the lifetime of a thin film depending on the applied strain amplitude ε_a, and the fraction of damage F. The introduction of the fraction of damage F as an additional parameter allows describing to what extend the thin film is damaged due to fatigue. This gives the opportunity to differentiate between different degrees of damage at the surface, which is especially useful for applications where optical properties are needed. For example, if a highly sensitive optical sensor is used, even 10 % damage fraction at the surface might mark the lifetime. Contrary to that, if the reflectivity is not that important even 50 % of damage at the surface may be tolerable, corresponding to longer a lifetime. In the following, a damage growth model will be introduced.

Considering the damage formation evolution, it can be treated as a limited exponential growth. This seems to be justified by the observed damage sequence with damage features that are nucleated followed by their growth. The following equation, similar to a law by Mitscherlich [122], is proposed to describe the damage formation:

$$F = F_{\max}\left(1 - \exp\left(-\frac{1}{N_{0.5}} \cdot N_F\right)\right)^n \qquad \textit{Eq. 6.1}$$

while F is the fraction of damage, $F_{\max}$ is the maximum fraction of damage, $N_{0.5}$ is the half lifetime, N_F is the number of cycles to a specific fraction of damage F, and n is the growth mode. Furthermore, basic assumptions of the model are that the growth mechanism of the damage features does not change, which is equivalent to a constant n. Also it was assumed that the features do not impinge each other and the damage formation does not continue after certain damage has been accumulated. To obtain a lifetime diagram calculated from experimental data, several steps have to be performed. All necessary information is incorporated in the reflectivity scans after a series of cyclic loading displayed in Fig. 6.4 (A). As demonstrated by the quantitative analysis of the thin films, the relative reflectivity is the opposite of the fraction of damage ($F = 1 -$ rel refl). Therefore, the relative reflectivity data at constant strain amplitudes for a certain cycle number can then be plotted as fraction of damage versus the cycle number (see Fig. 6.4 (B)). This data can be fitted with Eq. 6.1. $F_{\max}$ determines the upper boundary of the fit and is limited due to the fact that even fully damaged thin films retain a certain reflectivity. n is equivalent to the slope of the steady state region. Both, $F_{\max}$ and n are assumed to be constant within one sample. The half lifetime $N_{0.5}$, is the cycle number where half of the surface is damaged and depends on the strain amplitude. Assuming that a Basquin Law (Eq. 2.2) type equation applies for $N_{0.5}$, it can be rewritten as:

$$N_{0.5} = K \cdot \varepsilon_a^q \qquad \textit{Eq. 6.2}$$

by introducing two other fit parameters q and K. The parameter q can be calculated into the known fatigue sensitivity exponent b in the following way:

$$b = \frac{1}{q} \qquad \textit{Eq. 6.3}$$

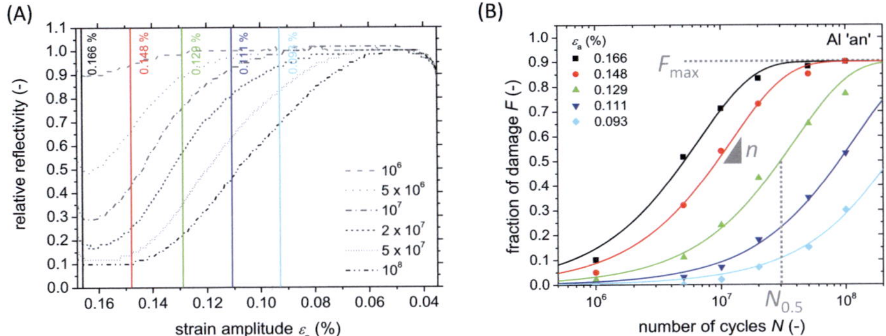

Fig. 6.4: (A) Reflectivity scans measured in the center of the cantilever for different cycle numbers, plotted versus the strain amplitude. The vertical lines indicate the values of the chosen strain amplitude for (B). Here, the fraction of damage is plotted versus the cycle number for different strain amplitudes and results in a growth curve, which can be fitted.

Using Eq. 6.1 and Eq. 6.2 and the obtained fit parameters the strain amplitude ε_a can be calculated depending on the cycle number for a certain fraction of damage in the following way:

$$\varepsilon_a = \left(-\frac{N_F}{K \cdot \ln\left(1 - \left(\frac{F}{F_{\max}}\right)^{\frac{1}{n}}\right)} \right)^{\frac{1}{q}} \qquad \textit{Eq. 6.4}$$

This gives the opportunity to derive a lifetime diagram with the strain amplitude versus the number of cycles to failure for different fractions of damage. The calculated lifetime diagrams for Al and Cu thin films are displayed in Fig. 6.5 and Fig. 6.6, respectively. Additionally the measured data obtained from the reflectivity scans (Fig. 5.10, Fig. 5.22) is plotted with open symbols for comparison. The fitted parameters are summarized in Table 6.3.

Table 6.3: Summary of calculated fit parameters of Al and Cu thin films.

sample	F_{max}	n	K	q	*b* (calculated)	*b* (measured)
Al 'as dep'	0.8	0.8	523	-5.9	-0.17	-0.14
Al 'an'	0.9	1.0	485	-5.5	-0.18	-0.11
Cu 'as dep'	0.7	0.4	3.6×10^{-10}	-23.6	-0.04	-0.04
Cu 'an'	0.8	0.5	0.116	-11.4	-0.09	-0.07
Cu 'on Ti'	0.7	0.5	0.015	-12.0	-0.08	-0.05

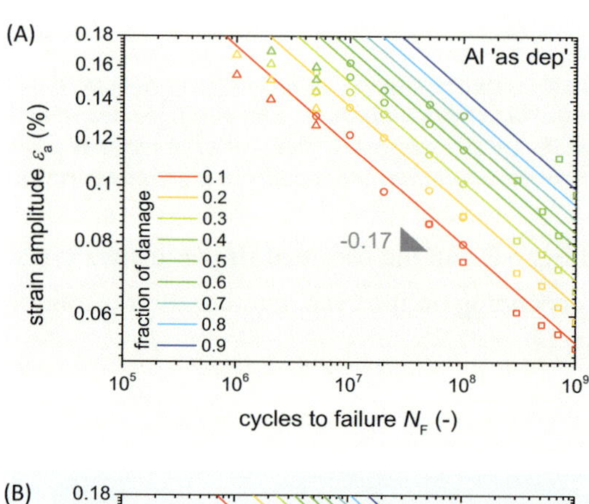

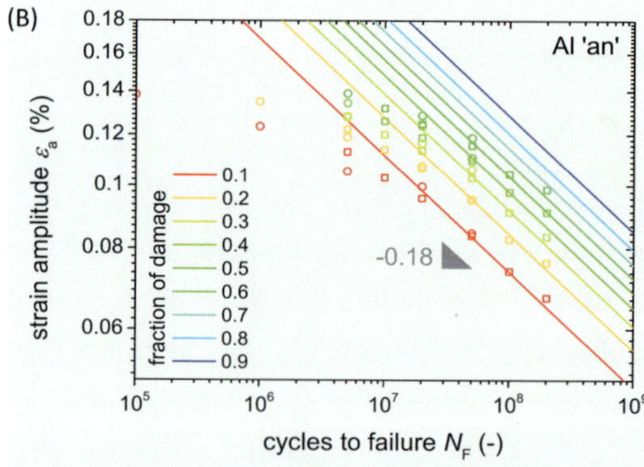

Fig. 6.5: Calculated lifetime diagram for different fractions of damage for (A) 'as deposited' and (B) 'annealed' Al thin films. The open symbols represent the measured data from reflectivity scans.

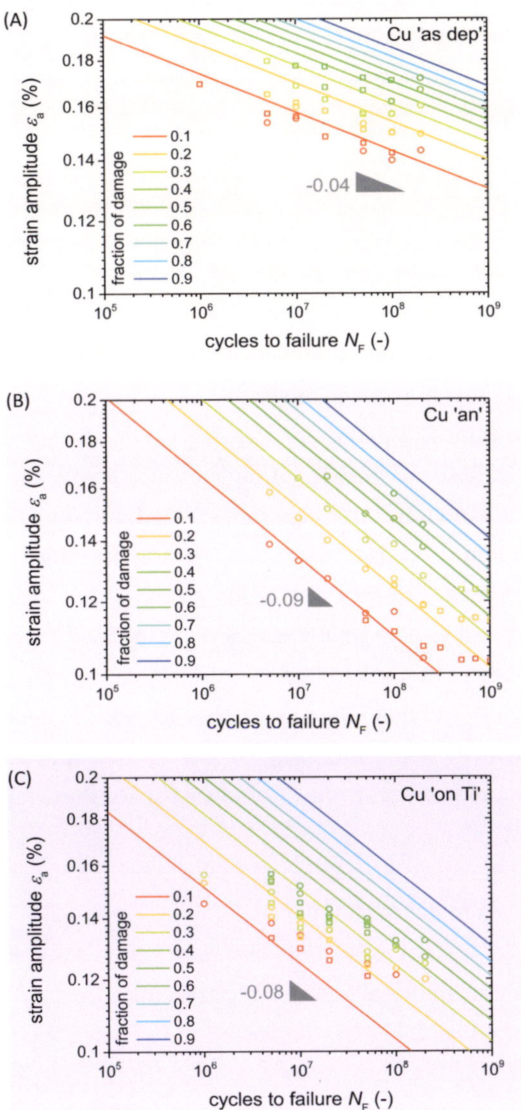

Fig. 6.6: Calculated lifetime diagram for different fractions of damage for (A) 'as deposited', (B) 'annealed', and (C) 'on Ti' Cu thin films. The open symbols represent the measured data from reflectivity scans.

In general, the model is in good agreement with the experimental data. However, some differences can be obtained. The calculated sensitivity factor b is overestimated for all the thin films besides Cu 'as deposited'. These differences might result in the fact that the model is only calculated with one sample for each material condition. If more data is used for fitting, the statistics would be better. In the case of the 'annealed' thin films there are deviations from the model, for Al at a lower cycle number and for Cu at a higher cycle number. This might be due to a change in the growth mode and therefore n seems not to be constant over the whole range of cycle numbers, which was a basic assumption for this model. In a next step of modeling, separate regimes of cycle numbers could be fitted with separate n, due to different damage modes.

For the Al thin films n is approximately 1, what can be attributed to homogenous damage nucleation and growth throughout the sample surface. This is also found for the microstructural investigations of Al thin films (chapter 5.2.1). In the case of the Cu thin films the growth mode n is in the range of 0.5. This might indicate that the damage formation depends on more than one mechanism and damage nucleation and growth might have a more inhomogeneous nature. Something similar has been obtained by the microstructural investigations of Cu thin films (chapter 5.3.1). The fact that n is smaller for the 'as deposited' thin films than for the 'annealed' thin films may be due to the influence of a different mechanism. This could be the cyclically induced grain growth prior to the damage formation.

The model has the most discrepancy for the Cu 'on Ti' thin films. Comparing the experimental data to the model, the deviation of the slope is fairly high. It has to be noted that the relative reflectivity displays a high slope depending on the position and therefore shows a rather sharp transition between undamaged and damaged region (see Fig. 5.17 (C)), compared to the other two Cu thin film conditions (Fig. 5.17 (A)+(B)). This makes it more difficult to acquire the necessary data for the fit. For further experiments the cycling steps have to be reduced to obtain more data points for the fitting curves. Besides the difficulties in data acquisition, the deviation of model and experimental data might be caused by an additional effect of the Ti seed layer and its influence on the damage formation mechanisms as was described before.

In conclusion, the model works most reliable in the range of 10^7 to 10^8 cycles, which may be something like a steady state region for damage formation. This is a regime between the LCF and the possible VHCF where the damage mechanism is rather stable. For a more realistic description, it has to be assumed that the damage mechanism can change and therefore n is not constant for the whole lifetime diagram. With this model a new way to describe the lifetime of a thin film additionally depending on the fraction of fatigue damage been proposed.

7. Summary

In this chapter the main findings of this thesis are concluded. It could be shown that the idea of using a strain amplitude gradient along a cantilever offers the opportunity to perform high-throughput fatigue testing of thin films. A resonant cantilever bending setup was developed with an *in situ* reflectivity measurement to analyze the damaged area. It was shown that these reflectivity scans directly correspond to the amount of damage caused by fatigue cycling. By defining a failure criterion correlating to a certain amount of damage, one lifetime diagram can be obtained with only one cantilever.

The characterization of the damage morphology for Al and Cu thin films with different initial grain sizes gave a good insight into how damage formation occurs in such thin films in the HCF and VHCF regime. In general, the Cu thin films display longer lifetimes than the Al thin films. Also, the fine-grained thin films have longer lifetimes as their coarse-grained counterparts. A special situation is the Cu 'on Ti' thin films, as the Ti seed layer has a strong influence on the fatigue damage. This allows engineering the fatigue properties. A damage mechanism sequence for each type of thin film was proposed.

For Al thin films a homogenously distributed damage formation is found, consisting mainly of hillocks and extrusions. The fine-grained 'as deposited' Al thin films first show grain growth, assumed to be induced by diffusion processes. Then, hillocks form by diffusion processes along the grain boundaries. Extrusions and pores form by dislocation glide on preferred glide planes. Short cracks occur between damaged areas to relief stresses. The coarse-grained 'annealed' Al thin films display hillock formation which is believed to be caused by grain rotation and sliding. Similar as before, extrusions form by dislocation processes. Pores could be induced in this case also by grain rotation. Short cracks may be caused by a local stress relief between damaged areas.

The Cu thin films only show extrusion formation as damage at the surface but are inhomogeneously distributed, which can be explained by the elastic anisotropy of Cu which induces local stress inhomogenities. The 'as deposited' Cu thin films show grain growth assumed to be mediated by stress driven grain boundary motion. Extrusions and pores lay on slip lines and might be caused by irreversible dislocation glide. Due to the annihilation of dislocations, vacancies are formed which can be accumulated to pores. The coarse-grained 'annealed' Cu thin films extrusion and pore formation therefore seems to be mediated by dislocation processes. Short cracks were observed near differently oriented extrusion islands to relief accumulated local stresses. The fine-grained Cu thin films deposited 'on Ti' seed layer to enhance the adhesion had a pronounced effect on the damage formation. Stress driven grain boundary migration might be the cause for the observed grain growth. The extrusion islands observed are larger than the others, which could be attributed to an in-plane texture of the grains. Furthermore, no pores have been found near the interface with the substrate. It is believed to be the reason because dislocations pile up at the interface and create a backstress and drive all the dislocation activity towards the surface, where an enhanced pore formation is observed. Short cracks occurred to be induced by stress relief near differently oriented extrusions.

In this thesis a novel high-throughput method was developed to test fatigue in thin films. It gives the opportunity to obtain one lifetime diagram with only one sample. Furthermore, it is possible to measure the amount of induced damage fraction. A phenomenological lifetime model has been proposed that introduces the fraction of damage as an additional variable, and opens the opportunity for a new kind of reliability estimation. In the future, it is possible to test a variety of different thin films, either pure element thin films or even, composites or multilayers. This technique opens the doors for systematic high-throughput testing of thin films.

8. References

1. W.A.J. Albert, *Über Treibseile am Harz.* Archive für Mineralogie, Geognosie, Bergbau und Hüttenkunde, 1838. **10**: p. 215-234.

2. A. Wöhler, *Versuche zur Ermittlung der auf die Eisenbahnwagen-Achsen einwirkenden Kräfte und der Widerstandsfähigkeit der Wagen Achsen.* Zeitschrift für Bauwesen, 1860. **10**.

3. L.F.J. Coffin, *A Study of the Effect of Cyclic Thermal Stresses on a Ductile Metal.* 1953.

4. S.S. Manson, *Behavior of materials under conditions of thermal stress.* National advisory commission on aeronautics. Vol. Report 1170. 1954.

5. O.H. Basquin, *The exponential law of endurance tests.* Proceedings of the ASTM, 1910. **10**: p. 625-630.

6. W.D. Nix, *Mechanical Properties of Thin Films.* Metallurgical Transactions a-Physical Metallurgy and Materials Science, 1989. **20**(11): p. 2217-2245.

7. D.T. Read and J.W. Dally, *Fatigue of Microlithographically-patterned Freestanding Aluminium Thin Film Under Axial Stresses.* Journal of Electronic Packaging, 1995. **117**(1): p. 1-6.

8. R.P. Feynman, *There's plenty of room at the bottom.* 1959.

9. C.-S. Han, H. Gao, Y. Huang and W.D. Nix, *Mechanism-based strain gradient crystal plasticity—II. Analysis.* Journal of the Mechanics and Physics of Solids, 2005. **53**(5): p. 1204-1222.

10. E. Arzt, *Overview no. 130 - Size effects in materials due to microstructural and dimensional constraints: A comparative review.* Acta Materialia, 1998. **46**(16): p. 5611-5626.

11. G. Dehm, C. Motz, C. Scheu, H. Clemens, P.H. Mayrhofer and C. Mitterer, *Mechanical size-effects in miniaturized and bulk materials.* Advanced Engineering Materials, 2006. **8**(11): p. 1033-1045.

12. K.J. Hemker and W.N. Sharpe, *Microscale characterization of mechanical properties.* Annual Review of Materials Research, 2007. **37**: p. 93-126.

13. O. Kraft, P.A. Gruber, R. Monig and D. Weygand, *Plasticity in Confined Dimensions.* Annual Review of Materials Research, Vol 40, 2010. **40**: p. 293-317.

14. J.R. Greer and J.T.M. De Hosson, *Plasticity in small-sized metallic systems: Intrinsic versus extrinsic size effect.* Progress in Materials Science, 2011. **56**(6): p. 654-724.

15. M. Ohring, *Materials Science of Thin Films*. 2nd ed. 2002: Academic Press.

16. S. Suresh and L.B. Freund, *Thin Film Materials: Stress, Defect Formation and Surface Evolution*. 1st ed. 2004: Cambridge University Press.

17. W.C. Oliver and G.M. Pharr, *An Improved Technique for Determining Hardness and Elastic-Modulus Using Load and Displacement Sensing Indentation Experiments.* Journal of Materials Research, 1992. **7**(6): p. 1564-1583.

18. D. Kiener, K. Durst, M. Rester and A.M. Minor, *Revealing deformation mechanisms with nanoindentation.* Jom, 2009. **61**(3): p. 14-23.

19. W.C. Oliver and G.M. Pharr, *Nanoindentation in materials research: Past, present, and future.* Mrs Bulletin, 2010. **35**(11): p. 897-907.

20. M.D. Uchic, P.A. Shade and D.M. Dimiduk, *Micro-compression testing of fcc metals: A selected overview of experiments and simulations.* Jom, 2009. **61**(3): p. 36-41.

21. O. Kraft, M. Hommel and E. Arzt, *X-ray diffraction as a tool to study the mechanical behaviour of thin films.* Materials Science and Engineering a-Structural Materials Properties Microstructure and Processing, 2000. **288**(2): p. 209-216.

22. R.D. Nyilas, S. Frank and R. Spolenak, *Revealing Plastic Deformation Mechanisms in Polycrystalline Thin Films with Synchrotron XRD.* Jom, 2010. **62**(12): p. 44-51.

23. H. Van Swygenhoven and S. Van Petegem, *The Use of Laue Microdiffraction to Study Small-scale Plasticity.* Jom, 2010. **62**(12): p. 36-43.

24. M. Legros, D.S. Gianola and C. Motz, *Quantitative In Situ Mechanical Testing in Electron Microscopes.* Mrs Bulletin, 2010. **35**(5): p. 354-360.

25. D.S. Gianola and C. Eberl, *Micro- and nanoscale tensile testing of materials.* Jom, 2009. **61**(3): p. 24-35.

26. M.A. Sutton, J.-J. Orteu and H. Schreier, *Image Correlation for Shape, Motion and Deformation Measurements*. 1st ed. 2009: Springer.

27. W.N. Sharpe, J. Pulskamp, D.S. Gianola, C. Eberl, R.G. Polcawich and R.J. Thompson, *Strain measurements of silicon dioxide microspecimens by digital imaging processing.* Experimental Mechanics, 2007. **47**(5): p. 649-658.

28. R.P. Vinci and J.J. Vlassak, *Mechanical behavior of thin films.* Annual Review of Materials Science, 1996. **26**: p. 431-462.

29. O. Kraft and C.A. Volkert, *Mechanical testing of thin films and small structures.* Advanced Engineering Materials, 2001. **3**(3): p. 99-110.

30. P. Chaudhari, *Plastic Properties of Polycrystalline Thin-Films on a Substrate.* Philosophical Magazine a-Physics of Condensed Matter Structure Defects and Mechanical Properties, 1979. **39**(4): p. 507-516.

31. M. Murakami, T.S. Kuan and I.A. Blech, *Mechanical Properties of Thin Films on Substrates.* London: Academic Press, 1982. **24**.

32. L.B. Freund, *The Stability of a Dislocation Threading a Strained Layer on a Substrate.* Journal of Applied Mechanics-Transactions of the Asme, 1987. **54**(3): p. 553-557.

33. M.F. Doerner and W.D. Nix, *Stresses and Deformation Processes in Thin-Films on Substrates.* Crc Critical Reviews in Solid State and Materials Sciences, 1988. **14**(3): p. 225-268.

34. E.O. Hall, *The Deformation and Ageing of Mild Steel .3. Discussion of Results.* Proceedings of the Physical Society of London Section B, 1951. **64**(381): p. 747-753.

35. N.J. Petch, *The Cleavage Strength of Polycrystals.* Journal of the Iron and Steel Institute, 1953. **174**(1): p. 25-28.

36. C.V. Thompson, *The Yield Stress of Polycrystalline Thin Films.* Journal of Materials Research, 1993. **8**(2): p. 237-238.

37. B. von Blanckenhagen, P. Gumbsch and E. Arzt, *Dislocation sources and the flow stress of polycrystalline thin metal films.* Philosophical Magazine Letters, 2003. **83**(1): p. 1-8.

38. R. Venkatraman and J.C. Bravman, *Separation of Film Thickness and Grain-Boundary Strengthening Effects in Al Thin-Films on Si.* Journal of Materials Research, 1992. **7**(8): p. 2040-2048.

39. R.M. Keller, S.P. Baker and E. Arzt, *Quantitative analysis of strengthening mechanisms in thin Cu films: Effects of film thickness, grain size, and passivation.* Journal of Materials Research, 1998. **13**(5): p. 1307-1317.

40. G. Dehm, T.J. Balk, H. Edongue and E. Arzt, *Small-scale plasticity in thin Cu and Al films.* Microelectronic Engineering, 2003. **70**(2-4): p. 412-424.

41. G. Dehm, T.J. Balk, B. von Blanckenhagen, P. Gumbsch and E. Arzt, *Dislocation dynamics in sub-micron confinement: recent progress in Cu thin film plasticity.* Zeitschrift Fur Metallkunde, 2002. **93**(5): p. 383-391.

42. T.J. Balk, G. Dehm and E. Arzt, *Parallel glide: unexpected dislocation motion parallel to the substrate in ultrathin copper films.* Acta Materialia, 2003. **51**(15): p. 4471-4485.

43. H. Gao, L. Zhang, W.D. Nix, C.V. Thompson and E. Arzt, *Crack-like grain-boundary diffusion wedges in thin metal films.* Acta Materialia, 1999. **47**(10): p. 2865-2878.

44. O. Kraft, L.B. Freund, R. Phillips and E. Arzt, *Dislocation plasticity in thin metal films.* Mrs Bulletin, 2002. **27**(1): p. 30-37.

45. R.S. Fertig and S.P. Baker, *Simulation of dislocations and strength in thin films: A review.* Progress in Materials Science, 2009. **54**(6): p. 874-908.

46. J. Weissmuller and J. Markmann, *Deforming nanocrystalline metals: New insights, new puzzles.* Advanced Engineering Materials, 2005. **7**(4): p. 202-207.

47. M.A. Meyers, A. Mishra and D.J. Benson, *Mechanical properties of nanocrystalline materials.* Progress in Materials Science, 2006. **51**(4): p. 427-556.

48. H. Van Swygenhoven, P.M. Derlet and A.G. Froseth, *Stacking fault energies and slip in nanocrystalline metals.* Nature Materials, 2004. **3**(6): p. 399-403.

49. K.S. Kumar, S. Suresh, M.F. Chisholm, J.A. Horton and P. Wang, *Deformation of electrodeposited nanocrystalline nickel.* Acta Materialia, 2003. **51**(2): p. 387-405.

50. Y.T. Zhu, X.L. Wu, X.Z. Liao, J. Narayan, L.J. Kecskes and S.N. Mathaudhu, *Dislocation-twin interactions in nanocrystalline fcc metals.* Acta Materialia, 2011. **59**(2): p. 812-821.

51. A.J. Haslam, D. Moldovan, V. Yamakov, D. Wolf, S.R. Phillpot and H. Gleiter, *Stress-enhanced grain growth in a nanocrystalline material by molecular-dynamics simulation.* Acta Materialia, 2003. **51**(7): p. 2097-2112.

52. Higgins, *Grain-Boundary Migration and Grain Growth.* 1974.

53. M. Winning, G. Gottstein and L.S. Shvindlerman, *Stress induced grain boundary motion.* Acta Materialia, 2001. **49**(2): p. 211-219.

54. J.W. Cahn, Y. Mishin and A. Suzuki, *Coupling grain boundary motion to shear deformation.* Acta Materialia, 2006. **54**(19): p. 4953-4975.

55. P.A. Gruber, J. Bohm, F. Onuseit, A. Wanner, R. Spolenak and E. Arzt, *Size effects on yield strength and strain hardening for ultra-thin Cu films with and without passivation: A study by synchrotron and bulge test techniques.* Acta Materialia, 2008. **56**(10): p. 2318-2335.

56. L. Nicola, E. Van der Giessen and A. Needleman, *Discrete dislocation analysis of size effects in thin films.* Journal of Applied Physics, 2003. **93**(10): p. 5920-5928.

57. P. Chaudhari, *Hilock growth in thin films.* Journal of Applied Physics, 1974. **45**(10): p. 4339-4346.

58. F.Y. Genin, *Effect of stress on grain-boundary motion in thin films.* Journal of Applied Physics, 1995. **77**(10): p. 5130-5137.

59. C.Y. Chang and R.W. Vook, *The effect of surface aluminum-oxide films on thermally induced hillock formation.* Thin Solid Films, 1993. **228**(1-2): p. 205-209.

60. D. Kim, B. Heiland, W.D. Nix, E. Arzt, M.D. Deal and J.D. Plummer, *Microstructure of thermal hillocks on blanket Al thin films.* Thin Solid Films, 2000. **371**(1-2): p. 278-282.

61. J.A. Nucci, A. Straub, E. Bischoff, E. Arzt and C.A. Volkert, *Growth of electromigration-induced hillocks in Al interconnects.* Journal of Materials Research, 2002. **17**(10): p. 2727-2735.

62. J. Schijve, *Fatigue of Structures and Materials*. 2nd ed. 2009: Springer.

63. S. Suresh, *Fatigue of Materials*. 2nd ed. 1998: Cambridge University Press.

64. S.E. Stanzl-Tschegg and B. Schonbauer, *Mechanisms of strain localization, crack initiation and fracture of polycrystalline copper in the VHCF regime.* International Journal of Fatigue, 2010. **32**(6): p. 886-893.

65. M. Zimmermann, *Diversity of damage evolution during cyclic loading at very high numbers of cycles.* International Materials Reviews, 2012. **57**(2): p. 73-91.

66. H. Mughrabi, *On the life-controlling microstructural fatigue mechanisms in ductile metals and alloys in the gigacycle regime.* Fatigue & Fracture of Engineering Materials & Structures, 1999. **22**(7): p. 633-641.

67. H. Mughrabi, *On 'multi-stage' fatigue life diagrams and the relevant life-controlling mechanisms in ultrahigh-cycle fatigue.* Fatigue & Fracture of Engineering Materials & Structures, 2002. **25**(8-9): p. 755-764.

68. J. Rösler, H. Harders and M. Bäker, *Mechanisches Verhalten der Werkstoffe.* 3rd ed. 2008: Vieweg+Teubner.

69. U. Essmann, U. Gosele and H. Mughrabi, *A Model of Extrusions and Intrusions in Fatigued Metals 1. Point-Defect Production and the Growth of Extrusions.* Philosophical Magazine a-Physics of Condensed Matter Structure Defects and Mechanical Properties, 1981. **44**(2): p. 405-426.

70. Z.F. Zhang and Z.G. Wang, *Grain boundary effects on cyclic deformation and fatigue damage.* Progress in Materials Science, 2008. **53**(7): p. 1025-1099.

71. H. Mughrabi and H.W. Hoppel, *Cyclic deformation and fatigue properties of very fine-grained metals and alloys.* International Journal of Fatigue, 2010. **32**(9): p. 1413-1427.

72. H.A. Padilla and B.L. Boyce, *A Review of Fatigue Behavior in Nanocrystalline Metals.* Experimental Mechanics, 2010. **50**(1): p. 5-23.

73. G.P. Zhang and Z.G. Wang, *Fatigue of Small-Scale Metal Materials: From Micro- to Nano-Scale* in *Multiscale Fatigue Crack Initiation and Propagation of Engineering Materials: Structural Integrity and Microstructural Worthiness*, G.C. Sih, Editor. 2008, Springer. p. 275-326.

74. D.T. Read, *Tension-tension fatigue of copper thin films.* International Journal of Fatigue, 1998. **20**(3): p. 203-209.

75. R. Schwaiger and O. Kraft, *High cycle fatigue of thin silver films investigated by dynamic microbeam deflection.* Scripta Materialia, 1999. **41**(8): p. 823-829.

76. R. Monig, R.R. Keller and C.A. Volkert, *Thermal fatigue testing of thin metal films.* Review of Scientific Instruments, 2004. **75**(11): p. 4997-5004.

77. S. Eve, N. Huber, O. Kraft, A. Last, D. Rabus and M. Schlagenhof, *Development and validation of an experimental setup for the biaxial fatigue testing of metal thin films.* Review of Scientific Instruments, 2006. **77**(10).

78. Y.C. Wang, T. Hoechbauer, J.G. Swadener, A. Misra, G. Hoagland and M. Nastasi, *Mechanical fatigue measurement via a vibrating cantilever beam for self-supported thin solid films.* Experimental Mechanics, 2006. **46**(4): p. 503-517.

79. C. Eberl, R. Spolenak, O. Kraft, F. Kubat, W. Ruile and E. Arzt, *Damage analysis in Al thin films fatigued at ultrahigh frequencies.* Journal of Applied Physics, 2006. **99**(11).

80. D. Wang, C.A. Volkert and O. Kraft, *Effect of length scale on fatigue life and damage formation in thin Cu films.* Materials Science and Engineering a-Structural Materials Properties Microstructure and Processing, 2008. **493**(1-2): p. 267-273.

81. X.J. Sun, C.C. Wang, J. Zhang, G. Liu, G.J. Zhang, X.D. Ding, G.P. Zhang and J. Sun, *Thickness dependent fatigue life at microcrack nucleation for metal thin films on flexible substrates.* Journal of Physics D-Applied Physics, 2008. **41**(19).

82. Y. Yang, B.I. Imasogie, S.M. Allameh, B. Boyce, K. Lian, J. Lou and W.O. Soboyejo, *Mechanisms of fatigue in LIGA ni MEMS thin films.* Materials Science and Engineering a-Structural Materials Properties Microstructure and Processing, 2007. **444**(1-2): p. 39-50.

83. Y.C. Wang, A. Misra and R.G. Hoagland, *Fatigue properties of nanoscale Cu/Nb multilayers.* Scripta Materialia, 2006. **54**(9): p. 1593-1598.

84. J.S. Bae, C.S. Oh, K.S. Park, S.K. Kim and H.J. Lee, *Development of a high cycle fatigue testing system and its application to thin aluminum film.* Engineering Fracture Mechanics, 2008. **75**(17): p. 4958-4964.

85. M. Hommel and O. Kraft, *Deformation behavior of thin copper films on deformable substrates.* Acta Materialia, 2001. **49**(19): p. 3935-3947.

86. G.D. Sim, Y. Hwangbo, H.H. Kim, S.B. Lee and J.J. Vlassak, *Fatigue of polymer-supported Ag thin films.* Scripta Materialia, 2012. **66**(11): p. 915-918.

87. S. Eve, N. Huber, A. Last and O. Kraft, *Fatigue behavior of thin Au and Al films on polycarbonate and polymethylmethacrylate for micro-optical components.* Thin Solid Films, 2009. **517**(8): p. 2702-2707.

88. B. Zhang, K.H. Sun, Y. Liu and G.P. Zhang, *On the length scale of cyclic strain localization in fine-grained copper films.* Philosophical Magazine Letters, 2010. **90**(1): p. 69-76.

89. G.P. Zhang, K.H. Sun, B. Zhang, J. Gong, C. Sun and Z.G. Wang, *Tensile and fatigue strength of ultrathin copper films.* Materials Science and Engineering a-Structural Materials Properties Microstructure and Processing, 2008. **483-84**: p. 387-390.

90. R. Schwaiger, G. Dehm and O. Kraft, *Cyclic deformation of polycrystalline Cu film.* Philosophical Magazine, 2003. **83**(6): p. 693-710.

91. G.P. Zhang, C.A. Volkert, R. Schwaiger, P. Wellner, E. Arzt and O. Kraft, *Length-scale-controlled fatigue mechanisms in thin copper films.* Acta Materialia, 2006. **54**(11): p. 3127-3139.

92. C. Eberl, R. Spolenak, E. Arzt, F. Kubat, A. Leidl, W. Ruile and O. Kraft, *Ultra high-cycle fatigue in pure Al thin films and line structures.* Materials Science and Engineering a-Structural Materials Properties Microstructure and Processing, 2006. **421**(1-2): p. 68-76.

93. R. Schwaiger and O. Kraft, *Size effects in the fatigue behavior of thin Ag films.* Acta Materialia, 2003. **51**(1): p. 195-206.

94. Y.B. Park, R. Monig and C.A. Volkert, *Thermal fatigue as a possible failure mechanism in copper interconnects.* Thin Solid Films, 2006. **504**(1-2): p. 321-324.

95. Y.B. Park, R. Monig and C.A. Volkert, *Frequency effect on thermal fatigue damage in Cu interconnects.* Thin Solid Films, 2007. **515**(6): p. 3253-3258.

96. G.P. Zhang, C.A. Volkert, R. Schwaiger, R. Monig and O. Kraft, *Fatigue and thermal fatigue damage analysis of thin metal films.* Microelectronics Reliability, 2007. **47**(12): p. 2007-2013.

97. C. Eberl, D. Courty, A. Walcker and O. Kraft, *Stress-gradient induced fatigue at ultra high frequencies in sub micron thin metal films.* International Journal of Materials Research, 2012. **103**(1): p. 80-86.

98. A.P. Boresi and R.J. Schmidt, *Advanced Mechanics of Materials*. 6th ed. 2003: John Wiley & Sons.

99. T. Straub, *private communication*. 2012.

100. S. Burger, B. Rupp, A. Ludwig, O. Kraft and C. Eberl, *Fatigue Testing of Thin Films.* Key Engineering Materials, 2011. **465**: p. 552-555.

101. M. Ziemann, *Experimentelle Untersuchung der Materialermüdung in zyklisch belasteten Kupferdünnschichten*, in *Fakultät für Physik*. 2012, Ludwig-Maximilians-Universität München: München.

102. R.E. Boroch, R. Muller-Fiedler, J. Bagdahn and P. Gumbsch, *High-cycle fatigue and strengthening in polycrystalline silicon.* Scripta Materialia, 2008. **59**(9): p. 936-940.

103. D.H. Alsem, C.L. Muhlstein, E.A. Stach and R.O. Ritchie, *Further considerations on the high-cycle fatigue of micron-scale polycrystalline silicon.* Scripta Materialia, 2008. **59**(9): p. 931-935.

104. Z.H. Cao, P.Y. Li, H.M. Lu, Y.L. Huang and X.K. Meng, *Thickness and grain size dependent mechanical properties of Cu films studied by nanoindentation tests.* Journal of Physics D-Applied Physics, 2009. **42**(6).

105. G. Dehm and E. Arzt, *In situ transmission electron microscopy study of dislocations in a polycrystalline Cu thin film constrained by a substrate.* Applied Physics Letters, 2000. **77**(8): p. 1126-1128.

106. C. Motz, T. Schoberl and R. Pippan, *Mechanical properties of micro-sized copper bending beams machined by the focused ion beam technique.* Acta Materialia, 2005. **53**(15): p. 4269-4279.

107. R. Spolenak, W.L. Brown, N. Tamura, A.A. MacDowell, R.S. Celestre, H.A. Padmore, B. Valek, J.C. Bravman, T. Marieb, H. Fujimoto, B.W. Batterman and J.R. Patel, *Local Plasticity of Al Thin Films as Revealed by X-Ray Microdiffraction.* Physical Review Letters, 2003. **90**(9): p. 096102.

108. M. Legros, K.J. Hemker, A. Gouldstone, S. Suresh, R.M. Keller-Flaig and E. Arzt, *Microstructural evolution in passivated Al films on Si substrates during thermal cycling.* Acta Materialia, 2002. **50**(13): p. 3435-3452.

109. S. Burger, C. Eberl, A. Siegel, A. Ludwig and O. Kraft, *A novel high-throughput fatigue testing method for metallic thin films.* Science and Technology of Advanced Materials, 2011. **12**(5).

110. S. Meister, M. Funk and J. Lohmiller. *Grain and particle analysis with line intersection method.* Mathworks 2012; Available from: http://www.mathworks.com/matlabcentral/fileexchange/35203-grain-and-particle-analysis-with-line-intersection-method.

111. L.A. Giannuzzi and F.A. Stevie, *Introduction to Focused Ion Beams - Instrumentation, Theory, Techniques and Practice*. 1st ed. 2005: Springer.

112. A. Strecker, U. Salzberger and J. Mayer, *Specimen preparation for transmission electron microscopy (TEM) - reliable method for cross sections and brittle materials* Praktische Metallographie, 1993. **30**(10): p. 482-485.

113. O. Kraft, R. Schwaiger and P. Wellner, *Fatigue in thin films: lifetime and damage formation.* Materials Science and Engineering a-Structural Materials Properties Microstructure and Processing, 2001. **319**: p. 919-923.

114. O. Kraft, P. Wellner, M. Hommel, R. Schwaiger and E. Arzt, *Fatigue behavior of polycrystalline thin copper films.* Zeitschrift Fur Metallkunde, 2002. **93**(5): p. 392-400.

115. R. Monig, *Thermal Fatigue of Cu Thin Films*, in *Max-Planck-Institut für Metallforschung*. 2005, Universität Stuttgart: Stuttgart.

116. S.R. Agnew and J.R. Weertman, *Cyclic softening of ultrafine grain copper.* Materials Science and Engineering a-Structural Materials Properties Microstructure and Processing, 1998. **244**(2): p. 145-153.

117. H.W. Höppel, M. Prell, L. May and M. Göken, *Influence of grain size and precipitates on the fatigue lives and deformation mechanisms in the VHCF-regime.* Procedia Engineering, 2010. **2**(1): p. 1025-1034.

118. D.S. Gianola, S. Van Petegem, M. Legros, S. Brandstetter, H. Van Swygenhoven and K.J. Hemker, *Stress-assisted discontinuous grain growth and its effect on the deformation behavior of nanocrystalline aluminum thin films.* Acta Materialia, 2006. **54**(8): p. 2253-2263.

119. D. Weiss, *Deformation Mechanisms in Pure and Alloyed Copper Films*, in *Max-Planck-Institut für Metallforschung*. 2000, Universität Stuttgart: Stuttgart.

120. L. Margulies, G. Winther and H.F. Poulsen, *In situ measurement of grain rotation during deformation of polycrystals.* Science, 2001. **291**(5512): p. 2392-2394.

121. A.W. Thompson and W.A. Backofen, *The effect of grain size on fatigue.* Acta Metallurgica, 1971. **19**(7): p. 597-606.

122. E.A. Mitscherlich, *Das Gesetz des Minimums und das Gesetz des abnehmenden Bodenertrags.* Landwirtschaftliche Jahrbücher, 1909. **38**: p. 537-552.

Schriftenreihe des Instituts für Angewandte Materialien

ISSN 2192-9963

Die Bände sind unter www.ksp.kit.edu als PDF frei verfügbar oder als Druckausgabe bestellbar.

Band 1 Prachai Norajitra
Divertor Development for a Future Fusion Power Plant. 2011
ISBN 978-3-86644-738-7

Band 2 Jürgen Prokop
Entwicklung von Spritzgießsonderverfahren zur Herstellung von Mikrobauteilen durch galvanische Replikation. 2011
ISBN 978-3-86644-755-4

Band 3 Theo Fett
New contributions to R-curves and bridging stresses – Applications of weight functions. 2012
ISBN 978-3-86644-836-0

Band 4 Jérôme Acker
Einfluss des Alkali/Niob-Verhältnisses und der Kupferdotierung auf das Sinterverhalten, die Strukturbildung und die Mikrostruktur von bleifreier Piezokeramik $(K_{0,5}Na_{0,5})NbO_3$. 2012
ISBN 978-3-86644-867-4

Band 5 Holger Schwaab
Nichtlineare Modellierung von Ferroelektrika unter Berücksichtigung der elektrischen Leitfähigkeit. 2012
ISBN 978-3-86644-869-8

Band 6 Christian Dethloff
Modeling of Helium Bubble Nucleation and Growth in Neutron Irradiated RAFM Steels. 2012
ISBN 978-3-86644-901-5

Band 7 Jens Reiser
Duktilisierung von Wolfram. Synthese, Analyse und Charakterisierung von Wolframlaminaten aus Wolframfolie. 2012
ISBN 978-3-86644-902-2

Band 8 Andreas Sedlmayr
Experimental Investigations of Deformation Pathways in Nanowires. 2012
ISBN 978-3-86644-905-3

Band 9 Matthias Friedrich Funk
Microstructural stability of nanostructured fcc metals during cyclic deformation and fatigue. 2012
ISBN 978-3-86644-918-3

Band 10 Maximilian Schwenk
Entwicklung und Validierung eines numerischen Simulationsmodells zur Beschreibung der induktiven Ein- und Zweifrequenzrandschichthärtung am Beispiel von vergütetem 42CrMo4. 2012
ISBN 978-3-86644-929-9

Band 11 Matthias Merzkirch
Verformungs- und Schädigungsverhalten der verbundstranggepressten, federstahldrahtverstärkten Aluminiumlegierung EN AW-6082. 2012
ISBN 978-3-86644-933-6

Band 12 Thilo Hammers
Wärmebehandlung und Recken von verbundstranggepressten Luftfahrtprofilen. 2013
ISBN 978-3-86644-947-3

Band 13 Jochen Lohmiller
Investigation of deformation mechanisms in nanocrystalline metals and alloys by in situ synchrotron X-ray diffraction. 2013
ISBN 978-3-86644-962-6

Band 14 Simone Schreijäg
Microstructure and Mechanical Behavior of Deep Drawing DC04 Steel at Different Length Scales. 2013
ISBN 978-3-86644-967-1

Band 15 Zhiming Chen
Modelling the plastic deformation of iron. 2013
ISBN 978-3-86644-968-8

Band 16 Abdullah Fatih Çetinel
Oberflächendefektausheilung und Festigkeitssteigerung von niederdruckspritzgegossenen Mikrobiegebalken aus Zirkoniumdioxid. 2013
ISBN 978-3-86644-976-3

Band 17 Thomas Weber
Entwicklung und Optimierung von gradierten Wolfram/ EUROFER97-Verbindungen für Divertorkomponenten. 2013
ISBN 978-3-86644-993-0

Band 18 Melanie Senn
Optimale Prozessführung mit merkmalsbasierter Zustandsverfolgung. 2013
ISBN 978-3-7315-0004-9

Band 19 Christian Mennerich
Phase-field modeling of multi-domain evolution in ferromagnetic shape memory alloys and of polycrystalline thin film growth. 2013
ISBN 978-3-7315-0009-4

Band 20 Spyridon Korres
On-Line Topographic Measurements of Lubricated Metallic Sliding Surfaces. 2013
ISBN 978-3-7315-0017-9

Band 21 Abhik Narayan Choudhury
Quantitative phase-field model for phase transformations in multi-component alloys. 2013
ISBN 978-3-7315-0020-9

Band 22 Oliver Ulrich
Isothermes und thermisch-mechanisches Ermüdungsverhalten von Verbundwerkstoffen mit Durchdringungsgefüge (Preform-MMCs). 2013
ISBN 978-3-7315-0024-7

Band 23 Sofie Burger
High Cycle Fatigue of Al and Cu Thin Films by a Novel High-Throughput Method. 2013
ISBN 978-3-7315-0025-4